S0-BIX-417

1989 Year Book of Developmental Biology

Joel M. Schindler

CRC PRESS, INC.
Boca Raton, Florida

Direct all inquiries to CRC Press, Inc., 2000 Corporate Blvd., N.W., Boca Raton, Florida, 33431.

International Standard Book Number 0-8493-3300-8
International Standard Serial Number 1042-8607

Printed in the United States

THE EDITOR

Joel Schindler received his B.Sc. degree in biology from the Hebrew University in Jerusalem, Israel in 1973 and his M.Sc. degree in biochemistry from the same institution in 1975. The following year, he returned to the United States with his doctoral mentor, Professor Maurice Sussman, to complete his doctoral studies. He was awarded his Ph.D. from the University of Pittsburgh in 1978.

From 1978 to 1981, Dr. Schindler was a postdoctoral research fellow at the Roche Institute of Molecular Biology in Nutley, New Jersey. During this time period, Dr. Schindler's research efforts focused on the study of mammalian embryogenesis. In particular, he initiated studies to investigate changes in gene expression during the peri-implantation period of mouse development. In addition, he was involved in a series of studies aimed at unraveling the mechanism of action of retinoids in inducing murine embryonal carcinoma cell differentiation.

Following his tenure at Roche, Dr. Schindler became an Assistant and subsequently, an Associate Professor in the Department of Anatomy and Cell Biology at the University of Cincinnati College of Medicine in Cincinnati, Ohio. In addition, he was a member of the Graduate Program in Developmental Biology at the Institute for Developmental Research, Children's Hospital Research Foundation, Cincinnati. Dr. Schindler participated in several team-taught courses to both graduate and medical students and was primarily responsible for the areas of cell differentiation and early embryo development. His research efforts remained focused on the regulation of gene expression during cell differentiation and specifically included defining the role of polyamines in regulating the differentiation of both murine and human embryonal carcinoma cells.

In 1987, Dr. Schindler accepted a position in the Genetics and Teratology Branch at the National Institute of Child Health and Human Development (NICHD), in Bethesda, Maryland. His current responsibilities include developing and overseeing NICHD-supported projects in the areas of basic developmental genetics and early embryo development. His unique position allows Dr. Schindler to closely monitor current progress and publications in the field of developmental biology.

Dr. Schindler has received several fellowships and awards; has been a Visiting Fellow at Macquarie University, New South Wales, Australia; and has served as both an editorial and grant reviewer for numerous journals and funding institutions. He is a member of the Society for Developmental Biology, Sigma Xi, the American Society of Cell Biology, the American Association for the Advancement of Science and the New York Academy of Sciences. He is the author or coauthor of scientific reports in numerous journals, books and symposia volumes. This CRC volume is Dr. Schindler's first major editorial venture.

Gary M. Wessel, Ph.D.
Research Associate, Department of Biochemistry and Molecular Biology, University of Texas, M. D. Anderson Cancer Center, Houston, Texas

JOURNALS REPRESENTED

Anatomy and Embryology
Cell
Development
Developmental Biology
EMBO Journal
Gene
Genes Development
Genetics
Journal of Cell Biology
Journal of Cellular Physiology
Journal of Molecular Biology
Journal of Neuroscience
Molecular and Cellular Biology
Nature
Neuron
Proceedings of the National Academy of Sciences of the United States of America
Science

TABLE OF CONTENTS

Introduction 1

1
Developmental Genetics 3

2
Developmental Gene Expression 29

3
Developmental Cell Biology 51

4
Maternal Controls, Cytoplasmic Determinants and Imprinting 71

5
Cell Interactions 85

6
Cell Lineage and Developmental Fate 131

7
Cytodifferentiation — Cell and Tissue Specific Gene Expression and Maintenance 145

8
Homeobox Genes 179

9
Morphogenesis and Pattern Formation 201

Author Index 223

Subject Index 231

INTRODUCTION

Developmental biology as a field has experienced explosive growth during the last decade. Technological advances in both molecular genetics and protein chemistry have allowed investigators to explore areas of developmental biology with new-found precision. As a result, the amount of detail that has been uncovered is enormous. Yet with all this new-found detail, the field is still evolving.

Developmental biology represents a merger of two previously independent disciplines, genetics and embryology. These two disciplines were not only conceptually distinct, but they also exploited different experimental systems for investigation. Geneticists were interested in inheritance, and the long-term implications of inheritance evolution. Most found the fruit fly or the mouse to be the best experimental vehicle to address their ideas. Embryologists concerned themselves more with events that take place within one generation. They relied more on experimental manipulation and observation using experimental specimens like sea urchins or frogs. It was only when scientists realized that genetic information could direct embryological events and that they could apply genetic approaches to studying embryological questions that developmental biology came of age.

Recently, molecular characterization has become the focus of developmental biology, but extensive biological groundwork was needed as a foundation upon which to build our molecular components. Our current understanding of eukaryotic development suggests that it is a complex series of regulatory circuitry. It is more than simply the accurate expression of certain genes. It involves patterns of gene regulation, precise roles of gene products in cellular events, and the interactions of multiple gene products. In addition, genomes are not all functionally equivalent and the genetic transmission of DNA includes both the transmission of biochemical factors and the chemical status of the DNA molecule itself. *Cis*- and *trans*-acting factors at the genomic level, diffusable substances, and local interactions at the cellular level all combine to orchestrate the events that are necessary to ensure the accurate completion of normal developmental processes.

This volume is an attempt to present the most significant literature contributions in the area of developmental biology of the past year. With entire journals devoted to this topic, as well as other journals containing numerous contributions to the field of developmental biology, the number of published articles reaches well into the thousands. Selecting those that are "most significant" was not a trivial task. Obviously, selections reflect the biases of the selectors, the members of the advisory board. A different group of scientists would no doubt select a set of articles with slightly different foci, but certainly of equal merit.

Since developmental biology has evolved so rapidly into such a large and diverse field it is often the case that scientists studying developmental genetics in one experimental system simply do not have the time to adequately explore the literature of developmental genetics for other, equally interesting experimental systems. Therefore, one of the goals of this volume is to expose you to areas of investigation related to your primary interest, but just distant enough to be overlooked in the literature you normally read. We attempted to include all major aspects of animal development. Plant development has not been included, primarily and admittedly due to the naivete of the scientific editor. Future volumes will hopefully reverse that shortcoming.

The organization of this volume is two tiered. First, the chapters are organized by topics starting with the "least complex" at the level of the genome and progressing through cellular events to "complex" aspects of development that focus on entire organ systems or an entire organism. Since we know how formidable the "simple" genome is, we are building on a fairly sophisticated baseline. As you will see, the assignment of individual literature articles to specific chapters is somewhat arbitrary and could arguably have been done differently. For example, where should an article on the tissue-specific gene expression of a cell surface molecule involved in cell-cell interactions that influence cell lineage determination be assigned?

Within chapters, articles are organized phylogenetically from the least complex (lower eukaryotes) to the most complex (mammals). It is appropriate that certain topics are represented by a larger number of articles involving a particular experimental system, but overall, most topics contain representative publications from all the major experimental organisms.

The chapters do contain certain unifying themes suggesting that certain molecules or events may be fundamentally important throughout animal development and that there are threads of molecular conformity directing development. On the other hand, there is a wonderful variety of options that different organisms can select for their own unique developmental program.

Finally, another goal of this volume is to remind you what a wonderfully broad field the study of developmental biology truly is. All-too-often we develop blinders and focus so narrowly that we lose perspective of how our little piece of the puzzle really fits into the entire picture. It is our hope that this volume will remove those blinders and remind you that while each instrument alone is important, the entire orchestra plays the symphony.

Developmental Genetics 1

INTRODUCTION

Developmental genetics encompasses both the classical genetic approach and the contemporary molecular biological approach to unraveling the underlying genetic mechanisms that direct developmental events. A central them in both avenues of investigation is isolating and characterizing mutant phenotypes.

Mutant phenotypes have consistently provided a window through which developmental biologists could gaze at the genetic basis of developmental processes. Such phenotypes could be used to study linkage analysis, inheritance patterns, and, most importantly from a developmental biologist's point of view, aberrant development. Molecular genetics has allowed us to determine the actual structure of mutant genes, compare them to the structure of wild type genes, and predict the nature of the gene products. While once we relied on spontaneously occurring mutations and later could randomly induce them biochemically, now we can generate mutations by directly manipulating the genome. Thus, technological developments have allowed us to open the window wider and gaze deeper into the fundamental genetic basis of development.

The articles identified in this chapter focus on the fundamental strength of investigating mutant phenotypes. The papers deal with some classic ways to investigate mutant phenotypes (genetic segregation; mosaic analysis) and indicate how genetics can, at times, be more revealing than biochemistry. Most importantly, this chapter includes articles that demonstrate the different ways that mutant phenotypes can be generated through molecular genetic manipulations. These methods include using: transposable elements, insertional mutagenesis, homologous recombination, antisense RNA, site-directed mutagenesis, and transfection. In addition to these methods, some of the articles describe the "construction" of transgenic mice using infection with retroviral vectors, direct microinjection of exogenous DNA, and the novel incorporation of engineered embryonic stem cells. Of particular interest is the fact that such manipulations not only can be used to generate aberrant phenotypes that lose function, but also can be used to correct the expression of a mutant phenotype, or allow a "normal" cell to gain an additional function. Such technology has important and far-reaching implications for the diagnosis and treatment of certain human diseases.

Finally, it should be noted that not all experimental systems are directly amenable to mutant analysis. While slime mold, fruit flies, and mice have large numbers of mutant phenotypes that can be analyzed, chickens and amphibia do not, so they are less useful for genetic analysis.

Characterization of a Timing Mutant of *Dictyostelium discoideum* Which Exhibits "High Frequency Switching"

D. R. Soll, L. Mitchell, B. Kraft, S. Alexander, R. Finney, and B. Varnum-Finney

Dev. Biol., 120, 25—37, 1987 1-1

The preaggregative period of *Dictyostelium discoideum* represents the transition between the vegetative state and the onset of multicellularity, and, hence, has been closely studied. This has resulted in the description of two rate-limiting components, or "developmental timers", during this period. The present work was undertaken in order to isolate mutants exhibiting selective alterations in the lengths of these two components.

The timing mutant FM-1 was derived by UV irradiation of the axenic strain RC-3 and was initially selected for altered colony morphology. This mutant is normal with respect to the sequence of growth and morphogenesis, fruiting body and spore formation, and the timing of recapitulation, erasure, and dedifferentiation. However, its preaggregative period is reduced by 3.5 to 4.5 h or approximately 50%. Reciprocal shift experiments demonstrated that the first rate-limiting component of this period was missing in FM-1.

The FM-1 mutation behaves as a single recessive allele mapping to linkage group II. An unexpected characteristic of this strain was its high level of phenotypic instability, reminiscent of the phenotypic switching systems in *Candida albicans.* Not only was FM-1 itself subject to rapid phenotypic switches, but variants derived from FM-1 also were capable of rapid reversion as well as switching to other morphological and timing phenotypes. The authors suggest that mobile genetic elements may be responsible for these phenomena.

In summary, this study resulted in the identification of a *D. discoideum* mutant which exhibits a specific alteration in the program of developmental timing, without a concomitant effect on the sequence of development or the final morphology or function of the fruiting body. A genetic analysis of the network of genes responsible for the timing program in this organism is now feasible.

♦A mutant is described that is specifically defective in developmental timing. The mutant, FM-1, has a shortened preaggregative period, but the remainder of morphogenesis is normal. The mutation that is responsible for this phenotype, *fstA1*, maps to linkage group II. This genetic analysis shows the individuality and uniqueness of the first rate-limiting component of the preaggregative timer which until now had been demonstrated by conditional experiments.

FM-1 also exhibits the phenomenon of rapid phenotypic switching. These new developmental phenotypes are heritable and may be the result of a transposable element. *Stephen Alexander*

Antisense RNA Inactivation of Myosin Heavy Chain Gene Expression in *Dictyostelium discoideum*
D. A. Knecht and W. F. Loomis
Science, 236, 1081—1086, 1987 1-2

Although the function of myosin in muscle cell contraction is well understood, the role of this protein in nonmuscle cells is less clear. In order to investigate its function in *Dictyostelium discoideum*, mutants were generated by antisense RNA inactivation of myosin gene expression.

The single copy myosin heavy chain gene (*mhcA*) was inserted into a transformation vector in reverse orientation with respect to an actin 6 promoter, which also directs transcription of the neomycin phosphotransferase gene used to select transformants. Control cells were either untransformed or transformed with vector alone.

In axenic media, transformed cells grow under selective conditions, but slowly (~ 24-h generation time) compared to control cells (10- to 12-h generation time). Their growth requires attachment to a surface, unlike control cells. More striking was the observation of up to 50% giant, multinucleate cells, up to 10 times larger in diameter than control cells and containing up to 50 nuclei per cell. Although the mutant cells did not aggregate or "stream" normally, it was surprising to find them motile, despite lacking myosin.

Transformants were severely inhibited in *mhcA* gene expression as judged by several criteria: (1) Northern blot analysis showed dramatic reduction in mhcRNA expression compared to control cells; (2) Western blot analysis using antibodies to several myosin heavy chain epitopes showed undetectable levels of myosin heavy chain protein; and (3) electrophoretic separation of cell extracts followed by silver staining showed only a single difference between transformants and control cells, in that the former were missing the abundant and easily identified 243-kDa myosin heavy chain protein band. Northern blot analysis confirmed transcription of the antisense mhcRNA in transformants but not control cells.

In the presence of a bacterial food source, the actin 6 promoter is not normally active. When transformed cells were grown on a bacterial food source, they contained reduced concentrations of antisense RNA and nearly normal amounts of myosin heavy chain protein, and they appeared normal in terms of developmental timing, morphology, and growth. When returned to growth in axenic medium, transformants reverted to mutant phenotype. This provides evidence that the mutant phenotype is due to antisense inactivation rather than to other mutagenic effects of transformation.

While myosin appears dispensable for generalized locomotion in this organism, it is apparently required for normal cytokinesis as well as for aggregation and morphogenesis. These cells, in which myosin heavy chain expression can be regulated, should be useful in further studies of nonmuscle myosin.

♦The function of myosin was investigated by antisense RNA inactivation of myosin gene expression. Transformed strains had consistent mutant phenotypes including (1) slow growth in axenic medium and the necessity for attachment to the culture dish; (2) large cell volume with multiple (up to 50) nuclei; (3) lack of myosin heavy chain as determined by the presence of myosin heavy chain protein and antigenic activity (< 0.5% of the wild-type level); (4) the lack of the myosin heavy chain mRNA and the presence of the antisense RNA; and (5) poor streaming and aggregate formation with no further morphogenesis. The data indicate that the lack of myosin impairs cytokinesis and also has an effect on cell aggregation and morphogenesis. However, the mutant cells are clearly motile. Video recordings of the mutant cells indicates that the presence of cells with single nuclei is due to illegitimate cytokinesis where a nucleus occasionally buds off in a section of the cell. Using different methods, De Lozanne and Spudich (*Science,* 236, 1086—1091) have produced cells lacking myosin, and these cells have virtually identical phenotypes to those made by antisense mutagenesis. These strains and other similar mutants will be very useful for further investigations in molecular cytology. *Stephen Alexander*

Disruption of the *Dictyostelium* Myosin Heavy Chain Gene by Homologous Recombination

A. De Lozanne and J. A. Spudich

Science, 236, 1086—1091, 1987 1-3

Gene targeting or transformation by homologous recombination has been successfully used in yeast, but many properties of animal cells are not exhibited by yeast. This report describes the gene disruption by homologous recombination of the myosin heavy chain gene (*mhcA*) into *Dictyostelium*, a eukaryote which serves as a simple model for studying many cellular processes.

A clone was constructed by deletion of a portion of the *mhcA* gene, such that the remaining fragment would encode a myosin fragment equivalent to muscle heavy meromysin (HMM). This clone was inserted into a transformation vector containing the neomycin phosphotransferase gene under control of the *Dictyostelium* actin 15 promoter, in a plasmid which was incapable of autonomous replication in *Dictyostelium*. Transformation of AX4 cells was followed by a nonselective growth period in which the transformants which had not integrated the plasmid reverted to G418 sensitivity and lost the plasmid.

Seventy percent of the transformants were phenotypically normal by all criteria, including restriction map of the *mhcA* gene, *mhcA* RNA size and levels, and myosin heavy chain protein concentration. However, 30% were large (to 10 times normal diameter) and multinucleated. Cytokinesis was not observed in these mutant cells during time-lapse video microscopy over a 48-h period in which control cells divided several times. These mutants exhibit many varieties of motility, and also aggregate, but these aggregates are incapable of further development. Time-lapse photography showed occasional "pinching off" of a bud of cytoplasm containing a nucleus, but only of cells firmly attached to the substratum. This perhaps explained the mode of growth of these cytokinesis-deficient cells which could grow only when attached to plates and not when in suspension culture.

These abnormal *hmm* cells contained a *mhcA* gene restriction map consistent with that predicted by homologous recombination with the HMM fragment. Southern analysis demonstrated results consistent with the integration of five to ten copies of the plasmid in tandem arrays; Northern analysis demonstrated the absence of the native *mhcA* mRNA and the presence of a new mRNA apparently transcribed from the integrated HMM fragment; Western analysis showed the *hmm* cells to lack the myosin heavy chain and instead express a smaller cross-reacting species of the expected size of the HMM-encoded protein.

Reversion of *hmm* cells occur when cells are grown on bacterial lawns. These cells are normal in appearance, size, and development, and have reverted to sensitivity to the selecting agent. The restriction maps of *mhcA* gene from these cells show restoration of the wild-type pattern, and Western blot analysis shows the presence of native myosin heavy chain, and the absence of the smaller cross-reacting protein, demonstrating that reversion is due to excision and loss of the transforming plasmid.

♦A transforming vector carrying the *Dictyostelium* heavy meromyosin (HMM) fragment of the myosin heavy chain gene was inserted by homologous recombination into the single genomic copy of the myosin gene. The transformed cells are large and multinucleate suggesting that there is a defect in cytokinesis. When the transformed cells are grown on bacterial lawns, sectors occasionally arise that appear to be normal mononucleate cells. Molecular analysis demonstrated that these are revertants that have lost the insertion and are now expressing the wild-type myosin molecule again and are no longer resistant to G418 that was used to select for transformation. The results indicate that only a single insertion had occurred in the initial transforming event. In addition, homologous transformation may require a relatively large homologous region of DNA as compared to yeast. This work opens the way for replacing the myosin gene with an *in vitro* mutagenized counterpart. The results described in this paper dramatically complement those of Knecht and Loomis (*Science*, 236, 1081—1086) who use antisense inhibition of the myosin gene to study the function of myosin in these cells. *Stephen Alexander*

Sea Urchin Actin Gene Linkages Determined by Genetic Segregation

J. E. Minor, J. J. Lee, R. J. Akhurst, P. S. Leahy, R. J. Britten, and E. H. Davidson

Dev. Biol., 122, 291—295, 1987 1-4

In the sea urchin *Strongylocentrotus purpuratus* it is known from molecular biological techniques that cytoplasmic actin genes CyI-CyIIa-CyIIb are linked in the order shown, as are genes CyIIIa-CyIIIb, but it is not known whether these two clusters are linked to each other or to the muscle actin gene M. This study was undertaken to investigate this question.

The method employed involved the analysis of segregation of restriction fragment length polymorphisms (RFLPs). Highly heterozygous males were mated to females, and the resulting progeny were raised through metamorphosis. Southern blot analysis of the DNA of the progeny revealed the segregation of the various parental markers. Linkage between genes is indicated if a particular RFLP for one gene always cosegregates with a particular RFLP for another gene.

With data from 60 individual sea urchins using five probes, the inferences were made that the two gene clusters are unlinked either to each other or to the M actin gene, since segregation of RFLPs occurred at random (cosegregation values of 54 to 65%). The recombination intervals (kilobase pairs per percent recombination) is unknown in the sea urchin, but if it were the same as the mouse, a sea urchin chromosome would average 20 cM. Since actin gene clusters linked by 20 or fewer cM would have been detected in this study, this may indicate that these two actin gene clusters and the M actin gene are on separate chromosomes.

♦This study uses genetic segregation of RFLPs to demonstrate the linkage patterns of the actin gene family of *S. purpuratus*. The work shows that two actin gene clusters, CyI-CyIIa-CyIIb and CyIIIa-CyIIIb, and the M actin gene are not closely linked since RFLPs associated with these clusters segregate at random with respect to one another. Based on an assumed recombination interval, the authors also argue that the two actin gene clusters and the M actin gene are probably each on different chromosomes. The work represents the first studies on linkage analysis of genetic markers in sea urchins. It demonstrates that in a reasonable time interval (approximately 5 months), appropriate crosses can be made to generate the offspring required for the analysis. Given the large number of sea urchin genes that have now been cloned, the procedures described here could provide a way of mapping sea urchin genes at distances not easily accessible by molecular cloning and also generate a linkage map of the *S. purpuratus* genome. *William H. Klein*

Mosaic Analysis of Two Genes that Affect Nervous System Structure in *Caenorhabditis elegans*

R. K. Herman
Genetics, 116, 377—388, 1987 1-5

Genetic mosaic analysis was used to determine the cell autonomous nature of two behavioral mutants known to affect neuronal structure in *Caenorhabditis elegans.* The mutation *mec-4 (e1611)* causes the death of six sensory neurons, the microtubule cells, which mediate the response to touch. The fate of two of these cells, PLML and PLMR, was studied in animals mosaic for the *mec-4* mutation. In animals with *unc-3* mutations movement is uncoordinated and ventral cord motor neurons are disorganized. This mutant was also observed in these mosaic animals.

Using mosaic analysis it was determined that the *mec-4* mutation exerts its killing effect in a cell-autonomous manner. None of the neurons that synapse or form gap junctions with PLML or PLMR were involved in their degeneration and death. Mosaic analysis indicated that *unc-3* expression is required only in motor neurons for wild-type development and function of neurons. Body muscle and hypodermis were not the primary focus of *unc-3* activity.

Mosaic analysis predicts that *mec-4* would be transcriptionally active in microtubule cells and that *unc-3* would be transcriptionally active in cord motor neurons. However, mosaic analysis provides a test for function, not expression, and does not eliminate the possibility of transcription of these genes in cells in which their activity is not required.

♦This paper illustrates the power of mosaic analysis in *C. elegans,* i.e., one can determine the precise cellular focus of action of a mutant gene and thereby gain considerable insight into how the product of that gene carries out its function. Dr. Herman pioneered mosaic analysis with *C. elegans* and has worked out the focus of action of many functionally and developmentally important genes in this organism (R. K. Herman, *Genetics,* 108, 165—180, 1984). *Joseph G. Culotti*

Cholinergic Receptor Mutants of the Nematode *Caenorhabditis elegans*

J. A. Lewis, J. S. Elmer, J. Skimming, S. McLafferty, J. Fleming, and T. McGee

J. Neurosci., 7, 3059—3071, 1987 1-6

Selection for resistance to the neurotoxic drug, levamisole, which acts directly on cholinergic receptors, can be used to identify potential cholinergic receptor mutants in *Caenorhabditis elegans.* These mutants were then assayed for specific binding to tritiated meta-aminolevamisole in the presence and absence of mecamylamine, which causes allosteric binding activation, to assay for loss of receptor function.

Levamisole resistance associated with receptor abnormality occurred with mutation of seven genes. Mutants of *unc-29, unc-50,* and *unc-74* were severely deficient in saturable binding activity. Mutants of *unc-38, unc-63,* and *lev-7* had abnormally high-affinity, saturable-binding activity, that could not be further activated by mecamylamine. Mutants of *lev-1* had variable receptor activity, depending on the allele, but failed to become allosterically activated by mecamylamine. All seven genes appeared to be required for a functional levamisole receptor.

Seven genes required for the function of the putative nematode muscle acetylcholine receptor have been identified. These genes probably include the structural genes of the receptor and may also include genes that affect developmental expression or processing.

♦This paper identifies seven genes required for normal function of the acetylcholine receptor in *C. elegans.* This is more genes than have been found by biochemical methods to be required for receptor function and suggests that more receptor components are likely to be discovered by a genetic approach than by a purely biochemical one. It will be interesting to see what the products of these genes are and whether they have vertebrate counterparts. *Joseph G. Culotti*

Activation of a Transposable Element in the Germ Line but not the Soma of *Caenorhabditis elegans*

J. Collins, B. Saari, and P. Anderson

Nature (London), 328, 726—728, 1987 1-7

The rate of transposition of many transposable elements is tightly regulated. Tc1, a *Caenorhabditis elegans* transposable element, is under strain-specific and tissue-specific control. In the Bergerac strain of *C. elegans,* this element is fairly active in the soma, but not in the germ line. This report describes the isolation of mutants, derived from the Bergerac strain, with increased frequency of Tc1 transposition in the germ line.

Strain TR445, containing a Tc1 element in the unc-54 myosin heavy chain gene, was mutagenized with ethylmethane sulfonate. The offspring that exhibited increased unc-54 reversion frequencies were selected. These eight stocks were named ***mutator.*** Mutator mutants had germ-line reversion frequencies, due to excision and transposition of the Tc1 element, that were 4 to 100× higher than the original strain. There was no effect on the somatic Tc1 excision rate.

Mutator mutants will allow the cloning of *C.* ***elegans*** genes using transposon tagging techniques. Mutator lines can also be used to generate spontaneous mutations in many different genes, through increased transposition of the Tc1 element to novel sites.

♦This paper describes the isolation of mutator strains that enhance the frequency of transposon TC1 transpositions in *C. elegans.* These strains have allowed the cloning of many mutationally identified genes of *C. elegans* by transposon tagging. They have probably been the most useful molecular biology tool in *C. elegans* to date, although this is likely to change as the molecular fingerprint map of the *C. elegans* genome (Coulson et al., *Proc. Natl. Acad. Sci. U.S.A.,* 83, 7821—7825, 1986) becomes correlated with the genetic map. *Joseph G. Culotti*

Transposon-Induced Deletions in *unc-22* of *C. elegans* Associated with Almost Normal Gene Activity

J. E. Kiff, D. G. Moerman, L. A. Schriefer, and R. H. Waterston
Nature (London), 331, 631—633, 1988 1-8

Revertants of a Tc1 transposon-induced mutation in unc-22 were examined at the molecular level in the mutator strain of ***Caenorhabditis elegans.*** Four of these revertants were complete, and four had very subtle mutant phenotypes.

Southern analysis demonstrated that three of the partial revertants had deletions of approximately 1.0, 1.2, and 2.0 kb. The rest of the revertants had a wild-type pattern on the Southern blot, indicating more precise excision of the Tc1 element. The unc-22 gene was cloned from each revertant and sequenced. Only one revertant had the wild-type sequence, one had a single base change, and three had very small insertions. The other three revertants maintained the reading frame, but contained large deletions in the coding region, as expected. By immunoblot, each of these revertants produced a protein that was smaller than the wild-type protein. In the wild-type protein, the deleted region contains ten copies of a repeated motif. Therefore, lost function of these units might be covered by the remaining functional repeats.

These results indicate that in proteins containing repeated domains, large deletions can occur with the retention of nearly normal functioning. Redundancy of function allows the loss of some of these units to be covered by the functioning of the remaining units. The imprecise excision of transposons may be an important source of genetic diversity in protein structure and in hereditary disease.

♦The results of this paper may explain why deletions in a gene may result in only mild malfunctions of its product *in vivo,* that is, if the deletions remove (in frame) only some of a number of repeated motifs, the remaining copies of this motif may be functionally sufficient. *Joseph G. Culotti*

Insertional Mutagenesis of the *Drosophila* Genome with Single P Elements

L. Cooley, R. Kelley, and A. Spradling

Science, 239, 1121—1128, 1988 1-9

This report describes the creation of a line of *Drosophila* carrying a modified P element (jumpstarter) that is an abundant source of transposase, but is, itself, stable (see Figure 1-9A). When this line is mated to a line containing a single marked and defective P element (mutator, Figure 1-9B), this second element is mobilized. The random transposition events of the mobilized mutator element are captured in stocks that do not contain jumpstarter and are therefore stable. Mutator contains a bacterial origin of replication and a selectable drug resistance marker, to facilitate cloning in bacteria.

This P element-mutagenesis method was used to generate 1300 stocks containing single P element insertions. The *rotated abdomen* locus was

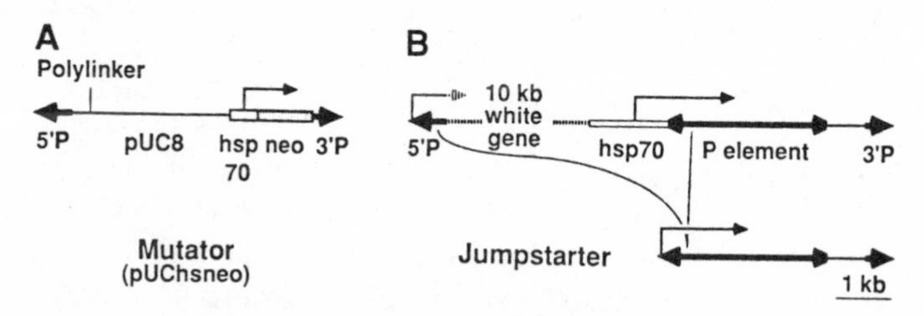

FIGURE 1-9. P element structure. (A) The mutator element, pUChsneo, contains the entire sequence of pUC8 including the *amp*^{R} gene and the ColE1 origin of replication, as well as a bacterial neomycin-resistance gene (*neo*^{R}) fused to a *Drosophila* heat shock protein promoter (*hsp70*). (B) The jumpstarter element was derived from the large transposon shown above, containing DNA that encodes both the *Drosophila white* gene, and a fusion of the *hsp70* promoter to the P element transposase gene. The *hsp70-P* fusion actually lies in the opposite orientation to the structure originally described. The jumpstarter element, shown below, is a derivative of this element in which all *white* and *hsp70* DNA is deleted. The thin solid line represents approximately 800 bp of DNA from pBR322 (Bal I to Sal I) that is present in the original white$^+$ transposon. The structure of this region in the jumpstarter element was not verified. The bracket indicates possible end points of the deletion, which appears to have resulted from a homologous recombination event. Transcription of the transposase gene in the jumpstarter element is assumed to be from the normal P element promoter (indicated by the bent arrow). (From *Science*, 239, 1123, 1988. Copyright by the AAAS. With permission.)

cloned from one of these strains to test the method. Approximately 10% of the P element insertions that were examined were lethal.

This method can be used to isolate mutants in any chromosome, generate secondary mutations at the site of insertion, and identify and clone previously undescribed genes with an interesting mutant phenotype. *Drosophila* lines containing single insertions will be an important resource for the molecular and genetic analysis of *Drosophila.*

A Stable Genomic Source of *P* Element Transposase in *Drosophila melanogaster*

H. M. Robertson, C. R. Preston, R. W. Phillis, D. M. Johnson-Schlitz, W. K. Benz, and W. R. Engels

Genetics, 118, 461—470, 1988 1-10

This report describes a P element that can mobilize other elements, but is itself stable. This element was found in a line that contained the P[*ry* + delta 2-3] element inserted in 99B.

Germ line mutability studies demonstrated that delta 2-3 (99B) produced as much transposase as a strong P strain. In a homozygous strain observed for 1 year, there was no transposition of this element, indicating extreme stability. This element was used successfully as a source of transposase for insertional mutagenesis studies, germ-line transformation studies, and mobilization of microinjected P elements.

These experiments have demonstrated that the delta 2-3 (99B) P element is a highly stable source of transposase. It can be used to increase the efficiency of the various techniques that require P element transposase, such as mutagenesis, transformation, and mobilization.

♦These two papers describe a new strain of flies carrying a modified P element (referred to as *delta 2-3* in this review) that is the source of abundant transposase activity. The element has the unusual property of being itself very stable, not being able to transpose in the genome.

The paper by Cooley et al. describes a method that uses the *delta 2-3* flies to identify and clone any *Drosophila* gene whose mutant phenotype can be recognized. A single defective "mutator" P element that is marked with a selectable marker is mobilized by the transposase produced by the immobile *delta 2-3* element (called "jumpstarter" by the authors). Random transpositions by the mutator are captured in individual stocks that no longer contain jumpstarter, where they remain stable. This method is far superior to the previous method of hybrid-dysgenic crosses because the latter crosses yield progeny containing a large number of P elements, making further analysis cumbersome and time consuming. The jumpstarter method yields mutants containing a single, stable P element, greatly facilitating the cloning of the gene. Since the mutator P element contains a bacterial origin of replication and a selectable drug resistance marker, the gene can be cloned by simply digesting mutant DNA with the appropriate restriction enzyme, circularizing it and directly transforming bacteria. The authors propose to use this method to create a "library" of fly stocks, each containing a single transposon in a defined point of the genome.

The paper by Robertson et al. also describes a *delta 2-3* strain of flies for which the authors give similar applications.

This method represents a quantal leap forward for the genetic analysis of *Drosophila,* giving it many of the powers of genetically marked tranposable elements in bacteria. *Marcelo Jacobs-Lorena*

Antisense RNA Inhibits Expression of Membrane Skeleton Protein 4.1 during Embryonic Development of Xenopus

D. H. Giebelhaus, D. W. Eib, and R. T. Moon

Cell, 53, 601—615, 1988 1-11

Membrane skeleton protein 4.1 is involved in modulating the binding of the membrane skeleton to the plasma membrane and the interaction of membrane proteins. To further examine the function of this protein, synthesis of protein 4.1 was perturbed in *Xenopus* embryos by microinjection of antisense RNA.

A cDNA encoding part of the chicken 4.1 gene was used to screen a lambda gt10 library for *Xenopus* 4.1 clones, which were then characterized. S1 nuclease protection assays using these clones, demonstrated that 4.1 transcripts remain at a constant level throughout *Xenopus* development. A partial clone of the 4.1 gene was inserted into the coding region of the CAT gene of the EMSV plasmid in both orientations. Microinjection of 100 pg of the antisense plasmid was lethal at the neurula stage, while microinjection of 100 pg of the sense plasmid had no effect. Microinjection of 50 to 75 pg of antisense plasmid was not lethal, permitting the level of 4.1 mRNA to be measured by S1 nuclease protection assays. After midblastula transition, no 4.1 mRNA could be detected in the embryos that received the antisense plasmid, although this protein could be detected in those embryos that received the sense plasmid. Embryos that received the antisense plasmid were smaller, developed more slowly, and rarely survived more than a few weeks. This phenotype could be rescued by the simultaneous injection of 20 pg of sense plasmid along with the antisense plasmid, indicating that the phenotype was due to the expression of the antisense plasmid. Rabbit antibodies to protein 4.1 were generated and used to monitor the expression of this protein in embryo tissues. Accumulation of 4.1 protein was reduced in the retina of antisense embryos, and the interphotoreceptor space was increased. Protein 4.1 was uniformly reduced in all tissues of the embryos expressing the antisense RNA.

The antisense RNA experiments indicated that protein 4.1 is essential for development, as embryos that received 100 pg of antisense RNA died at the neurula stage. Embryos that received 50 to 75 pg were smaller, had malformed organs, and died after a few weeks. In these embryos, protein 4.1 expression was uniformly reduced. In the retina of these embryos, the normal interaction of the outer segments of the photoreceptor with the pigment epithelia was perturbed by the lack of protein 4.1.

♦Substantial effort in several laboratories has gone into attempts to use antisense RNA to ablate the function of specific genes in *Xenopus* embryos, without success. This paper is the first case where is seems to have worked. The authors used an antibody specific for the protein they were trying to eliminate, and showed fairly clearly that it had been significantly reduced. There is one unsettled issue regarding this work, which is how the protein 4.1 antisense expression plasmid was able to affect most cells in the embryo; there are several unpublished reports circulating, including experience in my own laboratory, indicating that expression of DNA injected into fertilized *Xenopus* eggs is highly mosaic. If such were the case with the antiprotein 4.1 plasmid, then it is unclear how the observed result could have been obtained. *Thomas D. Sargent*

An Unstable Family of Large DNA Elements in the Center of the Mouse *t* Complex

J. Schimenti, L. Vold, D. Socolow, and L. M. Silver

J. Mol. Biol., 194, 583—594, 1987 1-12

A proximal part of mouse chromosome 17, the *t* complex, is found in two distinct forms within the genomes of wild populations of mice from around the world. One is a wild type while the other is a *t* haplotype. All *t* haplotypes are recent descendants of a single ancestral chromosome. These *t* haplotypes can express deleterious phenotypes, including male sterility and homozygous lethality, but *t*-carrying chromosomes are propagated at high frequency in the wild.

The authors cloned 363 kb from a novel, locally dispersed family of 11 large DNA elements, or T66 elements, within the center of mouse *t* haplotypes. Homologies among members of this family are noted along a repeated unit of at least 75 kb. The organization and number of elements in wild type forms of chromosome 17 were very different from those within *t* haplotype forms. Widely varying numbers of T66 elements were found within individual chromosomes in both inbred mouse strains and independently derived *t* haplotypes. While wild-type chromosomes had five to nine T66 elements distributed between two loci separated by at least three map units, *t* haplotypes had 9 to 11 T66 elements within a single cluster.

A set of testis-specific transcripts has been noted on analysis of Northern blots with probes derived from a region of the T66 family of elements. This supports the concept that sequences within the T66 region may function during germ-cell differentiation in male mice.

♦The *t* complex, which encompasses the proximal portion of mouse chromosome 17, exists in two forms. One form is considered to be wild-type while the other is called a *t* haplotype. Most wild populations of *Mus musculus* and *Mus domesticus* are polymorphic for different *t* haplotypes, each of which carries variant alleles of a number of genes that affect embryonic development, sperm formation, and function. Some of the phenotypic effects associated with each *t* haplotype include interaction with the dominant mutation *T* to produce taillessness in *T/t* animals, homozygous lethality, transmission ratio distortion in *+/t* or *T/t* males, and sterility in males homozygous for a semilethal *t* haplotype or doubly heterozygous for two different lethal *t* haplotypes. In addition to these effects, suppression of meiotic recombination has also been observed to occur in *+/t* heterozygotes, and this has been found to be due to two large inversions present within *t* haplotype chromosomes. Chromosome micro-dissection techniques have led to the establishment of a partial genomic library from which many cloned inserts have been identified as mapping to the *t* complex. Even though a significant amount has been learned in the past few years concerning the organization and molecular structure of the *t* complex, we are still as ignorant as ever about the nature of the defects attributed to the variant alleles associated with *t* haplotypes. The paper cited above is important in that data are presented concerning the cloning of a large region (approximately 400 kb) of the *t* complex known to contain the responder locus (involved in transmission ratio distortion). A locally dispersed family of eleven DNA elements (T66 elements, approximately 75 kb in length) arranged into three subfamilies were identified. Wild-type chromosomes were found to have between five and nine T66 elements distributed between two loci separated by 3 cm, whereas *t* haplotypes have between 9 to 11 T66 elements within a single cluster. These data suggest at least one inversion event has occurred in the wild-type ancestor chromosome but not in the *t* haplotype ancestor chromosome. Genetic studies have demonstrated the localization of the responder locus to the T66B subregion and a prominent set of T66 testes-specific transcripts have been identified. The authors have proposed that the T66 region may be involved in germ cell differentiation and possibly may be the responder locus itself. Although further work is needed to confirm these hypotheses, the possibility of having cloned a variant *t* haplotype gene is exciting. *Terry Magnuson*

Site-Directed Mutagenesis by Gene Targeting in Mouse Embryo-Derived Stem Cells

K. R. Thomas and M. R. Capecchi

Cell, 51, 503—512, 1987

1-13

It is possible to inactivate by gene targeting a specific locus in the mouse genome. The endogenous gene used as the target was hypoxanthine phosphoribosyl transferase (*Hprt*), which is on the X-chromosome. Selection procedures are available for isolating *Hprt* mutants. Embryo-derived stem cells were the recipient cell line. After inactivation of a selected gene, they should provide a means of generating mice with the desired mutation.

A construct of the neomycin resistance gene (*neo*^{r}) was introduced into an exon of a cloned fragment of the *Hprt* gene and used to transfect embryo-derived stem cells. One in 1000 G418′ colonies were resistant to the base analogue 6-thioguanine (6-TG). The G418′, 6-TG^r cells all were *Hprt* because of homologous recombination with the exogenous *neo*^{r}-containing *Hprt* sequences. Two classes of *neo*^{r}-*Hprt* recombinant vectors were compared: those replacing the endogenous sequence with the exogenous one and those inserting the exogenous sequence into the endogenous one. Targeting efficiencies of both types of vectors depended closely on the extent of homology between the exogenous and endogenous sequences.

This should prove to be a useful approach to targeting mutations into any gene. When embryo-derived stem cells are used as a host line and site-specific mutagenesis is achieved by gene targeting, mice of any desired genotype can be generated. The first-generation chimera usually will be heterozygous for the targeted mutation. Subsequent breeding will generate homozygous animals. In this way recessive lethals can be maintained as heterozygotes.

Targetted Correction of a Mutant HPRT Gene in Mouse Embryonic Stem Cells

T. Doetschman, R. G. Gregg, N. Maeda, M. L. Hooper, D. W. Melton, S. Thompson, and O. Smithies

Nature (London), 330, 576—578, 1987 1-14

Homologous recombination between a native target chromosomal gene and exogenous DNA can modify the target locus specifically in culture. This, along with mouse embryonic stem (ES) cells that can enter the germ line after genetic manipulation in culture, provides a means of making predetermined changes in mammalian germ lines.

The authors have used gene targeting to correct the mutant hypoxanthine-guanine phosphoribosyl transferase (HPRT) gene in an ES cell line previously used to produce an HPRT-deficient mouse. Three male ES cell lines with an inactive X-linked HPRT gene were used to produce mice that transmit the mutant genes to their progeny. The inactive HPRT gene is due to a spontaneous deletion or to integration of retroviral sequences into different introns of the gene. Human and mouse sequences were supplied to replace the deleted promoter. The correcting plasmid has 2.5 to 5 kb of DNA in common with the target locus. DNA analyses confirmed that the

HPRT⁻ mutation was corrected in the predicted way in HPRT⁺ colonies by plasmid sequences that integrated only into the target locus. The frequency of targeted correction was about 14% of that of obtaining G418-resistant colonies.

The authors presently are trying to generalize gene targeting by extending it to genes without selectable phenotypes and by avoiding introducing any foreign DNA into the target locus.

♦The work reported in these two papers represents a major breakthrough in the ability to manipulate the mouse genome. There are many examples of genes that have been cloned and are known to be expressed at critical times during mouse development. However, the function of these genes is not at all understood, primarily because no known mutations have been isolated that would result in an abnormal phenotype. Noteable examples of this category include the mouse homeobox genes and proto-oncogenes. The two reports listed above demonstrate that it is possible to achieve in embryo-derived stem cells site-specific alterations within a selectable gene *(Hprt),* and that this can occur at a high enough frequency (10^{-6} which can be reduced to 10^{-3} in conjunction with *Neo* selection) to warrant further work on nonselectable genes. In the first report (Thomas and Capecchi, 1988), recombinant vectors designed to introduce a mutation into exon 8 were used. The mutation frequency was found to be the same irrespective of whether the vector replaced a portion of the endogenous *Hprt* sequence with the homologous exogenous sequence, or instead, inserted an exogenous sequence into the gene. In both cases, the targeting efficiency was dependent on the extent of homology between the exogenous and endogenous sequences.

The second report (Doetschman et al., 1987) presents data using homologous recombination with exogenous DNA to correct a mutated *Hprt* gene in embryo-derived stem cells. The cause of the mutation was a spontaneous deletion of at least 10 kb that extended from somewhere 5¢ of the promoter into the second intron. The exogenous DNA introduced into the mutant cells included *Hprt* exons 1 to 3, and unlike the first report, plasmid sequences were not removed. The final size of the introduced DNA was 12.4 kb with only 2.5 to 5 kb being in common with the target locus. The precise length of homology was not determined because the exact endpoints of the endogenous deletion were not known. Five *Hprt*⁺ colonies were isolated under selective conditions and then expanded. Upon Southern blot analyses, all five colonies gave bands corresponding to genes that were correctly modified. The average frequency of functional correction was 1.4 ¥ 10^{-6} (without *Neo* selection). Unlike the first report, the smaller length of overlapping sequence homology between the introduced sequences and the endogenous gene did not significantly lower the recombination frequency.

The importance of these two reports is that they demonstrate the feasibility of obtaining specific, predetermined germline changes by homologous recombination in selectable genes of embryo-derived teratocarcinoma stem cells. What remains to be demonstrated is how this technique will work with nonselectable genes. The major problem at present is devising a scheme that will identify the 1 per 1,000,000 cells (or 1 per 1,000 if combined with *Neo* selection) that contain the correct site of integration in a nonselectable gene. *Terry Magnuson*

HPRT-Deficient (Lesch-Nyhan) Mouse Embryos Derived from Germline Colonization by Cultured Cells
M. Hooper, K. Hardy, A. Handyside, S. Hunter, and M. Monk
Nature (London), 326, 292—295, 1987 1-15

The properties of embryonic stem (ES) cells make it possible to use methods of somatic cell genetics to select cells with genetic modifications — such as recessive mutations — and introduce them into the mouse germ line. The authors selected variant ES cells deficient in hypoxanthine guanine phosphoribosyl transferase (HPRT) *in vitro* and used them to produce germ-line chimeras that resulted in female offspring heterozygous for HPRT deficiency. HPRT-deficient preimplantation embryos were derived from these females.

The ES cell line, E14, was derived from strain 129/Ola mouse blastocysts. Cells negative for HPRT were injected into host blastocysts. Of 34 male chimeras test-bred with females carrying appropriate markers, 19 have shown germ-line transmission of the marker derived from the HPRT- ES cells. The mutant allele was transmitted through the female germ line, and the resulting HPRT-deficient embryos survived to the morula stage. Half the male offspring were HPRT-deficient, indicating no significant loss of deficient male embryos *in utero*.

Human HPRT-deficient males have Lesch-Nyhan syndrome, characterized by mental retardation, spastic cerebral palsy, and self-mutilative biting. Lesions in the dopamine-mediated neural paths of the basal ganglia are presumed to be responsible. HPRT-deficient mice may serve to test this hypothesis, as well as to evaluate proposed treatments. The mice also might be used as a model system to rescue defective embryos by cell replacement or gene therapy.

A Potential Animal Model for Lesch-Nyhan Syndrome Through Introduction of HPRT Mutations into Mice
M. R. Kuehn, A. Bradley, E. J. Robertson, and M. J. Evans
Nature (London), 326, 295—298, 1987 1-16

Lesch-Nyhan syndrome is a rare disorder of males caused by an inherited deficiency of the purine salvage enzyme hypoxanthine-guanosine phosphoribosyl transferase (HPRT). It is not clear how the resulting changes in purine metabolism produce the severe symptoms of Lesch-Nyhan syndrome. Mutations at the *Hprt* locus leading to loss of enzyme activity have not been described in laboratory animals.

Cultured mouse embryonic stem cells mutagenized by retroviral insertion and selected for loss of HPRT activity were used to construct chimeric mice in the hope of obtaining an animal model of Lesch-Nyhan syndrome. Two clonal lines carrying different mutant *Hprt* alleles gave rise to germ cells in chimeras, permitting the derivation of strains of mutant mice having the biochemical defect of Lesch-Nyhan syndrome. Male mice carrying the mutant alleles were viable. Analysis of their cells showed HPRT activity to be absent.

As in humans, a lack of HPRT activity does not lead to prenatal mortality in mice. In analogy with humans, it may be some time before $HPRT^-$ mice exhibit gross abnormalities. This work suggests the possibility of deriving strains with specifically induced alterations in other genes. It may eventually be possible to produce specific changes in endogenous genes through homologous recombination with cloned copies modified *in vitro.*

♦Both reports demonstrate that it is possible to select for embryo-derived (ES) stem cells that exhibit an $HPRT^-$ phenotype, and then to use these cells in blastocyst-injection experiments to create chimeras that contain functional germ cell contribution derived from the injected stem cells. By breeding the chimeric mice, it was possible to establish animals that were deficient in HPRT enzyme activity. The same condition in humans results in Lesch Nyhan disease. Thus, these reports represent the first example of using mutant ES cells to create an animal model for a human disease. Because the structural locus for HPRT is X linked, male stem cell lines were used so that only a single mutational event was needed. In the first report (Hooper et al., 1987), spontaneous HPRT mutations were isolated (frequency of 10^{-6}) using *in vitro* selection techniques. In the second report (Kuehn et al., 1987), ES cells were first mutagenized by retroviral insertion and then selected for loss of HPRT activity (frequency of 10^{-7}). In one instance the provirus inserted into the 5′ end of the gene whereas in another it was in the 3′ end of the gene. In both cases the proviral insertion inactivated the endogenous HPRT gene. No significant loss of $HPRT^-$ male embryos *in utero* was reported in either papers, and no postnatal deaths had yet occurred. In fact, at the time of publication all $HPRT^-$ mice (ranging in ages 3 to 11 weeks) were healthy and free of any gross neurological symptoms. Thus, it remains to be determined whether the $HPRT^-$ mice will serve as a useful animal model for Lesch Nyhan disease. Nonetheless, the success of genetic manipulation of the ES cells and transferring these back into the embryo opens up numerous possibilities for deriving strains of mice carrying mutations in specific genes. *Terry Magnuson*

Inoculation of Newborn SWR/J Females with an Ecotropic Murine Leukemia Virus can Produce Transgenic Mice
J. J. Panthier, H. Condamine, and F. Jacob
Proc. Natl. Acad. Sci. U.S.A., 85, 1156—1160, 1988 1-17

Endogenous ecotropic murine leukemia proviruses not present in the parental stock are acquired by the progeny of some SWR/J × RF/J hybrid female mice. The authors made a stock of an ecotropic murine leukemia virus produced by such a hybrid female, and inoculated it into newborn SWR/J females. Three ecotropic proviruses are integrated into the chromosomes of RF/J mice, but are lacking in SWR/J mice.

On crossing inoculated females with SWR/J males, some of the progeny acquired ecotropic proviruses. Most of the proviruses appeared to be distributed in somatic tissues, but some are transmitted through the germ line. Exogenous infection therefore can mimic the phenomenon noted in SWR/J × RF/J hybrids, and this may occur during oogenesis in the recipient females. The ecotropic murine leukemia provirus derived from hybrid females was integrated in both germ-line and somatic cells of the embryos of SWR/J females inoculated at birth.

Mice transgenic for a replication-competent murine leukemia virus can be produced in a relatively straightforward way that does not require manipulation of eggs or early embryos. It may be possible to transfer any gene inserted into a defective murine leukemia viral vector using the same route.

♦The work reported here has expanded upon the original observations of Jenkins and Copeland (*Cell*, 43, 811—819, 1985) which demonstrated that the progeny of some SWR/J × RF/J F1 females carrying the tightly linked *Emv-16* and *Emv-17* ecotropic MuLV proviruses acquire new proviral insertion sites at a high rate (18%) in their somatic and germ line lineages. It appears that although SWR/J mice carry no endogenous ecotropic MuLV in their genome, they have a permissive genotype for viral production. The present report shows that newborn SWR/J females can be infected by intraperitoneal injection with the ecotropic MuLV particles originating from the SWR/J × RF/J F1 hybrids. *In situ* analysis revealed that at least two of the infected females showed viral expression in the ovaries. Eight of these inoculated females were bred and of the 129 progeny produced, 8% had acquired a total of 19 ecotropic proviruses (on average, 0.14 proviruses per animal). The newly acquired proviruses were present at less than one copy per cell indicating a mosaic distribution in the tissues. This would suggest that proviral insertion occurred after fertilization. Two of these females were tested and found to transmit the newly acquired proviruses to the next generation. Thus, it appears that the SWR-RF MuLV is capable of infecting the germ line, an observation that has not been observed with the Moloney MuLV. The important aspect of this report, besides the virology, is that this may represent an efficient means for beginning an insertional mutagenesis screen. The techniques for producing insertional mutations by microinjecting into the pronucleus of a fertilized egg or by infecting embryo-derived stem cells which are then selected and introduced into blastocysts by injection techniques are time consuming and laborious. Inoculation of newborn SWR/J females and analysis of progeny from these females by tail-blots for identification of new insertion sites should be less demanding. Given published frequencies, 0.4% of the newly acquired proviruses should result in a recessive lethal phenotype. If the frequency of newly acquired proviruses can be increased, this would certainly be the method of choice. *Terry Magnuson*

Perinatal Lethal Osteogenesis Imperfecta in Transgenic Mice Bearing an Engineered Mutant pro-α1(I) Collagen Gene
A. Stacey, J. Bateman, T. Choi, T. Mascara, W. Cole, and R. Jaenisch
Nature (London), 332, 131—136, 1988 1-18

Type I collagen is a major constituent of the extracellular matrix. Malfunction of type I collagen is observed in many forms of osteogenesis imperfecta. Substitutions of single glycine residues of α1 (I) collagen have been associated with osteogenesis imperfecta type II. Transgenic mice with a mutant α1 (I) collagen gene into which specific glycine substitutions are engineered exhibit a dominant lethal phenotype characteristic of human osteogenesis imperfecta.

Mutations were made in a mouse COLIAI genomic clone that resulted in substitution of a glycine residue at position 859 of the α1 (I) chain by a cysteine or an arginine. Mutant collagen RNA expression was detected in transfected fibroblasts. Impaired secretion of mutant collagen has been observed in some cases of lethal perinatal osteogenesis imperfecta. Microinjection of DNA from mutant cosmids into fertilized eggs showed that the mutant gene exerts a dominant lethal effect. Cells transfected with the mutant genes synthesized collagen with decreased electrophoretic mobility.

This is the first instance of generation of an animal model of a human genetic disorder by introducing defined alterations into a normal gene and transferring the mutant gene into transgenic animals. The effects of the mutation can be studied in normal animals carrying wild-type alleles. It is likely that mutations in critical regions of proteins forming multimeric structures will have dominant effects in transgenic animals. This should permit genetic analyses of the synthesis and assembly of collagens or cytoskeleton proteins, even if mutations in endogenous genes are unavailable.

♦Introduction of cloned genes into mice by microinjection techniques has been used by many groups to achieve integration of defined genes into the germ line. The purpose of these experiments has been to elucidate the DNA sequences involved in tissue and stage-specific gene regulation, and this has been done by attempting to achieve developmentally appropriate expression of foreign genes by correlating specific modifications in gene design with corresponding modifications in expression. Even though a significant amount of work with transgenic mice has been done with these goals in mind, more recent work has focused on the developmental consequences of aberrant gene expression in transgenic mice. The report summarized here represents one of the first examples of generating an animal model of a human genetic disease by introducing defined alterations into a normal gene and then producing transgenic mice with this altered construct. Using this approach, these investigators produced transgenic mice bearing a mutant $\alpha1$ (I) collagen gene (engineered by site-directed mutagenesis to contain substitution of a glycine residue at position 859 by either a cysteine or an arginine). The introduction of this altered gene into the genome of embryos produced a dominant lethal phenotype characteristic of the human disease osteogenesis imperfecta. Of the mice that survived birth, none carried the transgene, whereas seven of nine fetuses that died shortly before or after delivery carried between 0.25 to 6 copies of the mutated collagen gene. This lethality was not observed for any of the control genes used. More importantly, it appears that a small amount of mutant gene expression (ranging from 4 to 30% of the total collagen RNA) disrupts normal collagen production. For example, fetuses producing as little as 10% mutant RNA had only 46% type I collagen when compared with normal control levels. Although, it is not clear why as little as 10% mutant chains results in reduced amounts of normal type I leading to lethality, the investigators offer explanations that include mutant chains interfering with correct collagen alignment, fibril formation, or interactions. The types of problems being addressed by microinjection technology are expanding, and the ability to produce animals with dominant mutations will have important applications for understanding gene function during development. *Terry Magnuson*

Stable Expression of Immunoglobulin Gene V(D)J Recombinase Activity by Gene Transfer into 3T3 Fibroblasts

D. G. Schatz and D. Baltimore

Cell, 53, 107—115, 1988 1-19

Immunoglobulin chain genes as well as T-cell receptor genes are assembled in developing lymphocytes by somatic recombination of widely separated gene segments to form a complete variable region. This region is assembled from variable (V), joining (J), and in some cases diversity (D) gene segments in a highly regulated manner. Recombination may be at least partly regulated by accessibility of the DNA substrate. Little is known of the enzymatic mechanisms of V(D)J recombination.

The authors used gene transfer methods to stably confer on a fibroblast the ability to carry out V(D)J rearrangements. Retroviral recombination substrates were utilized. V(D)J recombinase activity was found only in pre-B cells. However, a DNA transfer protocol activated V(D)J recombination in an NIH 3T3 cell line. The recombinase activity was transferred in a further round of transfection, and then was expressed in fibroblasts at a level comparable to that in recombinationally active pre-B cells.

Transfer of a single active gene can stably confer on a fibroblast the activity to carry out V(D)J rearrangements. Probably expression of a single lymphoid-specific gene can confer V(D)J recombinase activity on the fibroblast. Induction of a lymphoid-specific function in a fibroblast suggests that attempts to select for other events of transdetermination might succeed, providing access to genes that control differentiated functions in various systems.

♦The development of antigen-responsive immunocompetent B and T lymphocytes from committed stem cells requires the rearrangement of the antigen receptor genes, a process of somatic recombination in which widely separated gene segments encoding variable (V), joining (J), and in some cases, diversity (D) segments of the receptors' variable domains are joined. The process is both highly ordered and regulated. The temporal programs of rearrangement and the recombination signals within the DNA are reasonably well understood, and it is believed that B and T cells share the same basic recombination mechanisms. Yet, little is known about the enzyme machinery which catalyzes the recombination events. The recent work of Schatz and Baltimore provides an elegant assay with which to measure recombination in transfected cells, a prerequisite for identifying genes encoding the essential components of this machinery. Retrovirus DNA recombination substrates were constructed so as to include human germ line variable region segments — Ig κ, V, and J segments in opposite transcriptional orientation — separated by a gene encoding a selectable marker and oriented such that inversional rearrangements are selected, and an independent and constitutively expressed gene for neomycin resistance to allow for the selection of cells that have integrated the retrovirus. The substrates were validated by rearranging when transfected into cells competent to catalyze rearrangement (recombination frequency about one event per 10^3 cell divisions). Moreover, when transfected into cells of the B lineage arrested at distinct stages of maturation, the substrates rearranged in only some of the cells, confirming the conclusion of Lieber et al. (*Genes Dev.*, 1, 751—761, 1987) that recombinase activity is expressed during a narrow window of B cell development. Substrate recombination did not occur in an NIH 3T3 cell line until it was "activated" by transfer of DNA from a human B cell lymphoma under conditions where each transfected cell should have received a single genomic fragment of about 40 to 50 kb, or about one gene. The mechanism of recombinase activation in the NIH

3T3 cells remains to be elucidated. Possible mechanisms include the transfection of the recombinase gene itself or of a gene which results in the reprogramming of the NIH 3T3 cells to express the components of the recombinase machinery. Nonetheless, the strategy promises to reveal the genes which control the development of differentiated lymphocyte functions.
Judith A. K. Harmony

Reconstitution of MHC Class I Specificity by Transfer of the T Cell Receptor and Lyt-2 Genes
J. Gabert, C. Langlet, R. Zamoyska, J. R. Parnes, A.-M. Schmitt-Verholst, and B. Malissen
Cell, 50, 545—554, 1987 1-20

The T-cell antigen receptor consists of two polypeptide chains, and each T-cell clone expresses a unique α/β chain combination that recognizes antigen in the context of cell surface molecules encoded by the major histocompatibility complex. T lymphocytes also have a set of surface molecules that enhance their avidity for target cells. These are LFA-1; CD2; CD8, or Lyt-2, 3 in mice; and CD4, or L3T4 in mice.

The authors transferred T-cell receptor α and β chain genes from an H-2 class I-specific, CD8-dependent cytotoxic T-cell clone, alone or combined with the Lyt-2 gene, into a class II-restricted, CD4$^+$ T-cell hybridoma. The α and β T-cell receptor genes lent the H-2 class I specificity of the donor to the recipient only if the same cell was transfected with the Lyt-2 gene. The functional Lyt-2 molecule was expressed on transfected cells in the absence of Lyt-3 polypeptide.

The Lyt-2 polypeptides can be normally or artificially expressed on the cell surface in the absence of the Lyt-3 subunit. Lyt-3 is not required for Lyt-2 to function in class I MHC recognition. Besides the T-cell receptor, Lyt-2 polypeptide is the only subset-specific molecule needed to retarget a class II-reactive, CD4$^+$ T-cell line toward H-2 class I molecules.

♦The T cell receptor for antigen recognizes antigen on antigen-presenting cells and only in the context of cell surface glycoproteins of the major histocompatibility complex, MHC. Mature immunocompetent human T cells are members of one of two predominant classes, CD4$^+$ (helper, effector) or CD8$^+$ (cytotoxic). The CD4 (L3T4 in mouse) and CD8 (Lyt-2,3 in mouse) membrane proteins belong to the immunoglobulin supergene family and are important in directing T cells to respond to antigenic challenge in the appropriate MHC background. Typically, CD4$^+$ cells recognize antigen presented in association with class II MHC molecules, whereas CD8$^+$ cells recognize antigen presented with class I MHC molecules. Thus CD4 and CD8 may recognize nonpolymorphic determinants on MHC molecules. CD4 and CD8 proteins have also been suggested to play a role in the transduction of signals in the cells that express them. In gene transfer experiments, Gabert et al. have directly established for the first time that, in addition to the α/β heterodimeric T cell receptor for antigen (TCR), TCR^+ cells must also express Lyt-2 for productive recognition of antigen in the context of class I MHC. The α and β genes of the TCR donated by an H-2 class I-specific ($H\text{-}2K^b$) CD8-dependent cytotoxic T cell clone were transfected, without or with the Lyt-2 gene, into a class II-reactive $L3T4^+$ Lyt-2,3^- T cell hybridoma. Cells transfected with α/β-TCR responded by producing interleukin-2 when stimulated with a nonspecific mitogen leukoagglutinin and with an anticlonotypic antibody which recognizes an epitope on the donor TCR contributed by both α and β chains. The latter result indicated that the transfected TCR was sufficient to generate responsiveness, given the appropriate and nonspecific stimulus. However, the transfected cells only responded to alloantigen $H\text{-}2K^b$ when they had been cotransfected with the Lyt-2 gene, and the response was inhibited by anti-Lyt-2.2 antibodies. The absence of a requirement for the second component of Lyt-2,3, the Lyt-3 protein, is puzzling although it is consistent with the fact that the cell surface CD8 antigen consists of homodimers of the Lyt-2 human homolog. In addition, it is intriguing that none of the transfected cell lines was capable of killing $H\text{-}2K^b$-positive EL4 cells, although the cells possessed specific receptors as well as a lytic machinery as evidenced by their ability to kill EL4 cells in the presence of lectins. The designer cells may therefore be useful to study the connection between engagement of the TCR and cytotoxicity. *Judith A. K. Harmony*

HLA-DQ_β Gene Contributes to Susceptibility and Resistance to Insulin-Dependent Diabetes Mellitus

J. A. Todd, J. I. Bell, and H. O. McDevitt

Nature (London), 329, 599—604, 1987 1-21

Insulin-dependent diabetes is polygenic in inheritance; the *HLA-D* region contributes an estimated 50% or more of the heritability. At least one susceptibility gene has been demonstrated to map to the MHC, most closely with the *HLA-D* region. The *HLA-DQ* genes, which are in linkage disequilibrium with *HLA-DR*, are more closely related to insulin-dependent diabetes than the *DR* genes. The authors sequenced the four major expressed polymorphic class II gene products isolated from three patients with insulin-dependent diabetes and several controls. A rapid method was used to produce adequate amounts of target gene sequences from RNA by complementary DNA synthesis and *in vitro* DNA amplification.

No unique class II sequences are found only in patients with insulin-dependent diabetes. However, the DQ β-chain amino-acid sequence correlates directly with susceptibility to insulin-dependent diabetes mellitus. The DQ-determined susceptibility is largely dependent on identity of amino acid residue 57 of the β-chain of this heterodimer. The correlation was supported by binding a unique Ser57 residue in the murine homologue of DQ_β, A_β isolated from a nonobese diabetic mouse strain (NOD), compared with other mouse A_β alleles.

DQ_β alleles that lack Asp at position 57 are significantly enriched in the Caucasian insulin-dependent diabetic population. It is not clear whether Asp-57 alone confers decreased susceptibility. However, DQ_β allelic polymorphisms, especially at position 57, probably determine the specificity and extent of the autoimmune response to islet-cell antigens through T-cell helper/suppressor activity. The requirement for homozygosity of Asp 57-negative DQ_β alleles could explain the recessive inheritance of MHC-linked susceptibility to insulin-dependent diabetes seen in man and in the NOD mouse. The *DR2-Dw2* DQ_β allele is unique in that a single gene dose protects strongly — though not totally — against insulin-dependent diabetes.

♦Certain DNA markers in the class II (HLA-D) region of the major histocompatibility complex (MHC) of humans segregate with the susceptibility to a number of diseases, notably autoimmune diseases such as rheumatoid arthritis and insulin-dependent diabetes mellitus (IDDM). IDDM, a disease which affects about 0.5% of Caucasians, is the consequence of the destruction of pancreatic beta cells which produce insulin. An autoimmune mechanism is indicated by the presence of T and B lymphocytes in the pancreas and by circulating anti-beta cell and anti-insulin antibodies in patients with IDDM. The HLA-D region contains three loci DP, DQ, and DR, and alleles of DR and DQ have been positively correlated with IDDM in population studies. Moreover, two alleles of DR have been negatively associated with IDDM, suggesting that resistance to the disease is also genetically conferred. Several DR alleles are in strong linkage disequilibrium with one of the DQ alleles. A particular DR-DQ combination is termed a haplotype. It has been estimated that the HLA-D region contributes more than 50% of the heritability of the disease, although no unique class II polymorphisms are found exclusively in IDDM subjects. Todd et al. sequenced the alleles of DR and DQ which are associated with both susceptibility and resistance. Analysis of the sequences revealed that only residue 57 of the DQ_β chain correlated with susceptibility and resistance to IDDM in Caucasian populations. The absence of aspartic acid in this position was statistically enriched in patients with IDDM. This result was strengthened by the finding of a unique ser-57 in the murine homolog of DQ_β, A_β isolated from the nonobese diabetic (NOD) mouse strain. Two asp-57-negative DQ_β alleles — particularly if contributed by a DR4 or DR3 haplotype — were strongly correlated with IDDM, possibly accounting for the apparent recessive inheritance of MHC-linked susceptibility in man and in the NOD mouse. In contrast, two asp-57-positive alleles were associated with almost complete resistance. The heterozygous state was associated with reduced risk of developing IDDM.

Thus, the DQ_{β} alleles appear to be susceptibility genes, although the DQ_{β} alleles do not explain all HLA-IDDM associations. The mechanisms responsible for these fascinating correlations remain unknown. Possible mechanisms include: the mediation by asp-57-positive DQ alleles of T cell suppression for the putative beta cell antigens or of elimination of potentially autoreactive T cells during T cell ontogeny. *Judith A. K. Harmony*

Developmental Gene Expression 2

INTRODUCTION

The temporal and spatial expression of certain genes is essential for the successful completion of programmed development. Such genes include structural genes that encode the necessary proteins for proper cellular function and regulatory genes that contribute to the cascade of informational signals that correctly orchestrate the expression of both structural genes and other regulatory genes.

As we learn which molecular events are necessary to produce accurate gene expression, it is becoming more apparent that developmental gene expression can be regulated at multiple levels.

The eukaryotic genome is extremely complex. The manner in which a single gene is regulated during development is related to the molecular organization of the genome and how that organization can influence gene expression. Structural features (promoters; enhancers) play a central role.

The articles identified in this chapter offer several examples of developmentally regulated genes. They represent a mere glimpse of the vast wealth of detail that is being accumulated about specific genes in specific organisms. Within different developmental contexts, gene expression will be explored further in several later chapters of this volume. For now, you are getting a brief peak at what developmental gene expression can encompass in several different experimental systems.

A Locus Regulating *N*-Acetylglucosaminidase Synthesis during Development of Dictyostelium

H. S. Judelson, R. A. Burns, and R. L. Dimond

Dev. Biol., 120, 170—176, 1987 2-1

Mutations in genes that control defined steps in the development of the cellular slime mold *Dictyostelium discoideum* eliminate the accumulation of development-controlled enzymes and prevent the expression of various morphologic phenotypes. Among the enzymes thought to be developmentally regulated is the acid hydrolase *N*-acetylglucosaminidase. Specific cellular activity of this lysosomal enzyme increased five- to tenfold during development. The enzyme is required for normal migration of an intermediate pseudoplasmodial stage.

The rate of *de novo* synthesis of *N*-acetylglucosaminidase has now been shown to parallel the pattern of its cellular accumulation. A monoclonal antibody that specifically recognizes two forms of the enzyme was utilized. A new locus, *nagC*, regulates enzyme accumulation, and mutations at *nagC* specifically prevent its accumulation. However, they do not influence the expression of other developmental phenotypes including the accumulation of other developmentally regulated enzymes. Enzyme accumulation is prevented by countering an increase in the relative synthetic rate of *N*-acetylglucosaminidase. The *nagC* mutation is recessive and maps to linkage group VI; it is unrelated to the structural gene for *N*-acetylglucosaminidase.

Structural enzyme gene mutants and mutants that misregulate general aspects of differentiation have previously been identified in this slime mold. However, *nagC* is the first gene identified which specifically regulates a singel enzyme without affecting other developmental processes.

♦The paper explores the relationship of *N*-acetylglucosaminidase to development and the molecular mechanism of accumulation of the enzyme during development. Using immunoprecipitation techniques, it was shown that the increase in enzyme specific activity is due to an increase in specific enzyme synthesis. Two bands of 67 and 68 kDa are synthesized but their relationship is unclear. The bands do not interconvert in pulse chase experiments which indicates that there is no precursor-product relationship. In addition, the immunoprecipitation of both molecular forms is inhibited by purified 68-kDa enzyme. It is suggested that there may be some subtle difference in post-translational processing. Three mutant strains were isolated which showed decreased accumulation of *N*-acetylglucosaminidase activity during development. One of these appears to be a regulatory gene mutation. The strain HMW604, does not complement with a previously isolated structural gene mutation *nagA*. However, the new mutation is recessive to *nagA*$^+$. The new regulatory gene mutation, *nagC*, maps to linkage group VI. The mutant strain makes, but does not respond to, an extracellular factor that stimulates cells in suspension to express *N*-acetylglucosaminidase. *Stephen Alexander*

Effect of Protein Synthesis Inhibition of Gene Expression during Early Development of *Dictyostelium discoideum*

C. K. Singleton, S. S. Manning, and Y. Feng

Mol. Cell. Biol., 8, 10—16, 1988 2-2

The authors identified several genes that are deactivated when the development of *Dictyostelium discoideum* begins, through differential screening of various cDNA libraries. These genes have in common a reduction in steady-state levels of the corresponding mRNAs at the onset of development and as development continues. The induction of such transiently expressed genes may be independent of protein synthesis and may represent a primary event in the initiation of development of the slime mold.

When development proceeded in the absence of protein synthesis, by inhibition with cycloheximide, the decrease in mRNA levels for most genes was normal or slightly accelerated. For about 5% of them, however, cycloheximide appeared to induce expression, as reflected by an increase in mRNA levels. This resulted from inhibition of protein synthesis, not cycloheximide itself. An enhanced transcription rate appeared to be responsible. Normal rates of transcription for these genes depend on continued protein synthesis during vegetative growth and development.

There are two general regulatory classes for deactivation of gene expression on initiation of the development of *D. discoideum*, one dependent on and one independent of protein synthesis. There are differences in the timing of changes in expression for genes in these classes. Genes in different subclasses may utilize factors that are distinct but act by the same overall mechanism.

♦A family of genes (H genes) were cloned and identified that are expressed in vegetatively growing cells. The expression of these genes is stopped with the onset of development which is initiated by starvation. Interestingly, in the presence of the protein synthesis inhibitor cycloheximide, the expression of these genes continues or is increased. It appears that there are two very different mechanisms for the deactivation of gene expression at the start of development. Other inhibitors of protein synthesis, puromycin and pactamycin, with different modes of action were shown to have the same effect on the deactivation of the H genes as did cycloheximide. Thus, the observed change in gene expression appears to be the result of the inhibition of protein synthesis rather than some secondary effect of the drug. Heat shock did not bring about a similar alteration of gene expression, suggesting that stress is not the general underlying mechanism of the effect. Continued protein synthesis is necessary for the proper expression of the H genes. Even after 8 h of normal development when H gene expression is low or absent, the switch to cycloheximide resulted in an induction of the expression of the H genes. Thus, the H genes showed increased transcription in the presence of, but not in the absence of, cycloheximide. Aggregationless mutants were examined with respect to the expression of the H genes. Some mutants were found that had reduced vegetative levels of H gene expression. However, cycloheximide could induce the H gene expression, indicating that the block in these cells was not irreversible. *Stephen Alexander*

Structure of Two Developmentally Regulated *Dictyostelium discoideum* Ubiquitin Genes

R. Giorda and H. L. Ennis

Mol. Cell. Biol., 6, 2097—2103, 1987 2-3

Previous work has identified proteins that are developmentally regulated and specific for spore germination in *Dictyostelium discoideum.* A number of mRNAs that are specifically expressed during germination have been cloned. RNA blot analysis showed that mRNA specific for the cDNA clone pLK229 is present in very low concentration through the life cycle of *D. discoideum,* and that its concentration rises markedly in spores and germinating spores.

The clone pLK229 was used to screen two genomic libraries, and two genomic sequences homologous to the clone were isolated and sequenced. Genomic clone p229 is identical to pLK229 and codes for a 381-amino acid polypeptide, which consists of five tandem repeats of the same 76-amino acid sequence. Clone λ 229 codes for a 229-amino acid protein containing three tandem repeats of the identical 76-amino acid sequence. The 76-amino acid repeat was found to be identical to human and bovine ubiquitin — execpt for two amino acid differences.

Ubiquitin is found in all eukaryotic cells, either free or bound to cell proteins. More ubiquitin genes exist in *D. discoideum* than have yet been sequenced. Various ubiquitin mRNA species are expressed during *Dictyostelium* spore development. The fact that ubiquitin mRNA is developmentally regulated indicates that ubiquitin functions in development. Ubiquitin may be required for the degradation of proteins unique to spore germination.

♦A number of mRNAs and proteins are specifically expressed during spore germination in *Dictyostelium.* Using a cDNA clone, pLK229, it was shown that the cognate message is in very low concentration throughout growth and development but increases dramatically in spores and germinating spores. Southern analysis reveals that there are at least six genomic fragments that have homology to the pLK229 cDNA. The sequences predicted for the protein products of these genes are identical to that of the 76 amino acid human ubiquitin protein except for substitutions at two positions (glycine for proline at position 19 and asparagine for threonine at position 22). The repeats themselves differ in up to 34 of the 228 bases. One additional amino acid is added to the last repeat of each gene (asparagine in one case and leucine in the other). Developmental regulation of ubiquitin gene has been reported in other species as well. Its developmental regulation argues for a specific role for ubiquitin in development. *Stephen Alexander*

Mutually Exclusive Interaction of the CCAAT-Binding Factor and of a Displacement Protein with Overlapping Sequences of a Histone Gene Promoter

A. Barberis, G. Superti-Furga, and M. Busslinger

Cell, 50, 347—359, 1987 2-4

The CCAAT sequence is found in either orientation upstream of many mRNA-coding genes of eukaryotes. The histone genes of the sea urchin are subject to close developmental and tissue-specific regulation, in contrast to all core histone genes of vertebrates that have been examined. The sperm histone H2B-1 gene of the sea urchin contains two octamer sequences and two CCAAT sequences upstream of its TATA box. Nuclear extracts were made from testis and from blastula and gastrula embryos in order to identify sequence-specific factors related to tissue-specific expression of the sperm H2B gene.

The CCAAT-binding factors in nuclear extracts of testis and blastula and gastrula embryos were not distinguished by mobility shift or methylation interference analysis. However, a testis-specific octamer-binding factor was identified in addition to the ubiquitous form. In DNAse I protection studies, only the CCAAT-binding factor of the testis extract interacted with the sperm H2B promoter. In the embryonic extracts, a novel factor bound with high affinity to sequences overlapping the proximal CCAAT element, preventing DNA interaction of the CCAAT-binding factor in the embryo.

The CCAAT displacement protein may act to repress sperm H2B gene transcription. The protein resembles bacterial repressors and also SV40 large T antigen, in that it interferes with the DNA binding of a positive transcription factor. The CCAAT displacement factor may be the postulated repressor of the developmentally regulated gamma-globin genes in vertebrates.

♦The elegant *in vitro* DNA protein binding studies described in this article provide a convincing explanation of why the sperm specific H2B gene is not expressed in embryogenesis. Evidence also is presented for a testes-specific nuclear protein which may be responsible, in part, for transcription of the sperm-specific H2B gene in spermatocytes. The authors suggest that both positive and negative control elements play important roles in the regulation of this gene. Their hypothesis is that the TATA, octamer, and CCAAT sequences and their respective DNA-binding proteins are responsible for transcriptional activity of the sperm H2B gene in spermatocytes. However, in embryonic extracts a novel factor, the CCAAT displacement protein, prevents the embryonic CCAAT binding protein from interacting with the sperm H2B promoter region. The evidence presented is based on DNA footprint analysis which can distinguish the CCAAT displacement protein from the CCAAT binding protein. The CCAAT binding proteins can only bind to the H2B promoter in the presence of excess specific oligonucleotide competitor of the CCAAT displacement protein. The CCAAT displacement protein binds preferentially even when the ratio of CCAAT binding protein is increased, and prebound CCAAT binding protein can be displaced by the CCAAT displacement protein. The CCAAT displacement protein binds to sequences overlapping the proximal CCAAT element and the authors make the reasonable suggestion that this embryonic protein is a transcription repressor. It is not clear why the distal CCAAT element is not a site for a CCAAT binding protein in embryonic extracts but the authors suggest that cooperative interactions may be interrupted by the displacement protein. The proof for the repressor hypothesis will come when these sequences are tested directly for transcriptional activity by mutating the appropriate sites and relieving the negative regulation in microinjected oocytes. This work has far-reaching implications for differential gene expression in embryogenesis. It is possible that the temporal and spatial control of several sea urchin genes which are currently under investigation could be the result, at least in part, of negative controlling elements. *William H. Klein*

Post-Transcriptional Restriction of Gene Expression in Sea Urchin Interspecies Hybrid Embryos
R. A. Conlon, F. Tufaro, and B. P. Brandhorst
Genes Dev., 1, 337—346, 1987 2-5

Synthesis of many paternal species-specific proteins is reduced in sea urchin interspecies hybrid embryos because of a lesser amount of some paternal mRNA species in these embryos. The missing paternal mRNAs might normally be persistent maternal mRNAs not requiring replacement by accumulating zygotic transcripts in embryos of the paternal species. Alternately, the mRNAs may accumulate normally in embryos after fertilization, while their accumulation is aberrant in hybrid embryos.

Cloned cDNAs were used to analyze the expression of several *Lytechinus pictus* genes in hybrid embryos. The cDNAs that were specific for paternal RNA sequences were present in only 2 to 20% of normal amounts in hybrid embryos derived from a cross of *Strongylocentrotus purpuratus* eggs with *L. pictus* sperm. Several of the RNA species were barely detectable in eggs, but accumulated up to 40-fold during *L. pictus* embryogenesis. Accumulation of some *L. pictus* transcripts also was reduced in the reciprocal cross. The transcriptional activity of a gene encoding a metallothionein was as great in hybrid as in homospecific embryos, but in hybrid embryos the cytoplasmic transcript accumulated to only 2 to 15% of the normal level.

The restricted expression of paternal genes in hybrid embryos does not reflect the persistence of stable maternal mRNA species in eggs. The restricted expression of some genes in hybrid embryos might be due to competition between gene or transcript homologues for factors needed in active expression. Further studies of this phenomenon might provide insight into the role of selective gene regulation during embryogenesis.

♦Brandhorst and co-workers have reinvestigated the classical species hybrid experiments in echinoderms which provided evidence that the maternal genome is largely responsible for directing events in the early stages of embryogenesis. In earlier studies using two-dimensional gel analysis of newly synthesized proteins, they made the surprising observation that most paternal species-specific proteins were not detectably synthesized in hybrid embryos, even though the paternal DNA appeared intact. These results left the conclusions of the earlier species hybrid experiments open to different interpretations since a critical assumption, that the paternal genes are expressed normally, is not valid. The results also raised interesting questions as to the nature of the restricted expression. In the present article, the authors first ask whether the restricted expression is due to persistence of maternal mRNA. In other words, proteins showing paternal suppression may be translated from stable maternal mRNAs in homospecific embryos and hence the genes encoding them would be inactive during embryogenesis. That this is not the case is shown by the identification of four mRNAs which show significant accumulation in homospecific embryos (i.e., there is active embryonic expression from these genes) but whose levels are dramatically restricted in species hybrids. The studies further show that the restricted expression is not due to maternal dominance since the opposite cross also shows low transcript accumulation. In one example, a metallothionein mRNA, nuclear run-on experiments show that the restricted expression is not at the level of transcription. The basis of the restricted expression is not known, but the authors favor the idea that mRNA stability is being effected. They propose a model based on autoregulation of protein synthesis where co-translational competition of nascent chains for translational factors occurs between the paternal and maternal mRNAs. The less competitive mRNA (it could be either the maternal or paternal species) is destabilized. Whatever the basis of the restricted expression, further investigation of the phenomenon is likely to provide useful insights into post-translational regulation in embryogenesis.
William H. Klein

Both Basal and Ontogenic Promoter Elements Affect the Timing and Level of Expression of a Sea Urchin H1 Gene during Early Embryogenesis
A.-C. Lai, R. Maxson, and G. Childs
Genes Dev., 2, 173—183, 1988 2-6

Several families of histone genes in sea urchins are expressed in a unique stage-specific manner in embryos and adult tissues. The authors cloned and sequenced the *Stongylocentrotus purpuratus* late H1-β gene, which is regulated coordinately with the H1-γ gene and shares much sequence homology with the 350 bp upstream of the mRNA initiation site. A series of deletion and point mutants of the H1-β gene were microinjected into fertilized sea urchin eggs, and expression of the cloned DNA was monitored during embryogenesis.

At least two distinct components of the promoter were found to be responsible for determining the time of maximal transcript accumulation. In the wild-type promoter, an upstream sequence element, USE IV, activated basal promoter activity at the blastula stage of development. All late H1 genes also contained a highly conserved GC-rich sequence resembling a low-affinity binding site for the mammalian transcription factor Sp 1. This factor is required for basal expression of the H1-β gene at all stages of embryogenesis. When the GC-rich sequence was converted to a perfect core Sp1 sequence, the H1-β transcripts accumulated much more, and peak accumulation shifted to the early blastula stage.

Both these gene-specific activator sequences may be important in determining temporal patterns of gene expression in early embryogenesis.

♦There are several important conclusions that can be drawn from this article. First, four upstream sequence elements, USE I, II, III, and IV, found as conserved sequences upstream of many histone genes, are shown to effect transcription of a late H1 histone gene when mutated and analyzed by the sea urchin microinjection expression system. Second, when the most upstream element is deleted (USE IV at –318 to –287) basal expression is observed but proper induction or accumulation of transcripts at the late blastula stage does not occur. Late histone genes are known to be activated at this time. Third, late H1 promoters contain an imperfect core SPI sequence in the USE I region (GGGCTG), whereas the early H1 promoters contain a perfect core SPI sequence (GGGCGG). A single mutation in the late H1 gene promoter converting the GGGCTG to GGGCGG dramatically changes the temporal specificity of transcription and makes the late H1 gene behave like an early H1 gene. This finding is of significance because the early to late histone gene switch is an important model system for studying differential gene expression in embryogenesis and the mechanisms involved in downregulating the early histone genes and upregulating the late histone genes are still unknown. In this regard, the authors suggest an interesting model to explain their data. Late H1 genes normally require the sea urchin SPI equivalent for transcription but because the SPI recognition site on the late H1 promoter is imperfect, the gene competes poorly with other genes for the SPI factor. This results in a low transcription rate. Late H1 transcripts accumulate to higher levels in late blastula stage embryos at least in part because of the activation of a factor which can recognize the USE IV element. Early H1 gene promoters contain a strong SPI binding site and thus are transcribed in early development at high levels.

This model does not explain the downregulation of the early histone genes at the blastula stage. *William H. Klein*

Developmental Regulation of Micro-Injected Histone Genes in Sea Urchin Embryos

L. Vitelli, I. Kemler, B. Lauber, M. L. Birnsteil, and M. Busslinger

Dev. Biol., 127, 54—63, 1988 2-7

Histone genes of the sea urchin are found in several distinct families that are differentially regulated in embryonic and adult tissues. Early, late, and sperm histone genes of *Psammechinus miliaris* have been cloned and characterized. The authors developed a DNA transfer system based on the mouse egg injection technique to study the developmental behavior of cloned histone genes of *P. miliaris.* Two related sea urchin species were injected, followed by fertilization of the eggs.

All five early histone genes studied were expressed in early blastula embryos. The expression rate of the early H2A gene was estimated at five- to tenfold lower. Transcripts of this gene accumulated during cleavage stages and decayed in late embryos, in parallel with endogenous early H2A mRNA. An introduced late H2B gene, in contrast, was incorrectly regulated. Its mRNA level did not increase from the blastula to the gastrula stage. The sperm H2B-1 gene, normally inactive during development, was 80 times less well expressed than the early H2A gene in transformed blastulas. A fusion gene with the early H2A promoter linked to the structural sperm H2B gene was efficiently transcribed.

Sea urchin embryogenesis is characterized by the cell lineage-specific expression of all genes analyzed to date in late embryos. The low level of expression of injected sperm H2B gene in early embryos may reflect a negative regulatory influence of a displacement protein, or absence of a testis-specific octamer-binding factor.

♦A major objective of this article and a related article by this group (Barberis et al., *Cell,* 50, 347—359, 1987) is to understand why the early histone genes are actively transcribed during early embryogenesis while the closely related sperm specific histone genes are silent. In the present series of experiments, the authors use the sea urchin microinjection expression system to ask whether exogenously introduced early and late embryonic histone genes and sperm-specific histone genes behave as their endogenous counterparts. Early H2A transcripts appear as predicted but unlike other reports with microinjected late histone genes (Lai et al., *Genes Dev.,* 2, 173—183, 1988; Colin et al., *Proc. Natl. Acad. Sci. U.S.A.,* 85, 507—510, 1988), a late H2B gene does not follow proper temporal expression. Rather, transcripts do not increase in abundance during the transition from the blastula to early gastrula stage. The explanation for this is unknown but the authors suggest that some late histone genes may be cell cycle regulated and when microinjected into eggs they may be situated mostly in nondividing cells and hence inactive. More importantly, the sperm-specific H2B gene is very inefficiently expressed in embryos thereby allowing the authors to construct and test a chimeric gene consisting of bases –415 to +55 of the early H2A gene and bases +9 to +755 of the sperm-specific H2B gene. This chimeric gene clearly behaves like an early histone gene, demonstrating that all the sequences required for early H2A gene expression are in the –415 to +55 base region and that the +9 to +755 base region of the sperm-specific H2B gene does not contain negative response elements. This work, coupled with other investigations using the sea urchin expression system (Lai et al., op. cit., 1988; Colin et al., op. cit., 1988) provide the foundation for elucidating the elements required in *cis* for the early/late/adult histone gene switch during the sea urchin life cycle. *William H. Klein*

Closely Linked Early and Late Histone H2B Genes are Differentially Expressed after Microinjection into Sea Urchin Zygotes

A. M. Colin, T. L. Catlin, S. H. Kidson, and R. Maxson
Proc. Natl. Acad. Sci. U.S.A., 85, 507—510, 1988 2-8

Little is known about gene regulation in the earliest phases of development. The histone genes of the sea urchin provide an excellent model for studying regulatory mechanisms of this type. An attempt was made to learn whether cloned early and late histone genes are expressed with proper developmental timing when injected into sea urchin eggs. The early and late H2B genes were injected together on the same plasmid so that observed differences in timing of expression could not be ascribed to differences in copy number arising after injection.

Early and late histone H2B genes from the sea urchin *Strongylocentrotus prupuratus* were linked in a single plasmid and injected into eggs of *Lytechinus pictus.* Monitoring of transcript levels by a RNAse protection assay showed that transcripts of both injected and endogenous early genes peaked in the blastula stage. Transcripts of the injected late gene were detected at this stage and increased subsequently, until at least the early gastrula stage, 28 h after fertilization.

The DNA sequences regulating the temporal patterns of early and late H2B gene expression appear to lie within the cloned DNA segments. The expression of injected early and late H2B genes is very similar to that of the endogenous genes.

♦As with other similar studies (Vitelli et al., *Dev. Biol.*, 127, 54—63, 1988; Lai et al., *Genes Dev.*, 2, 173—183, 1988) the authors here demonstrate

that early and late histone genes microinjected into sea urchin eggs behave appropriately with respect to time of transcript accumulation during embryogenesis. In this study, the early and late H2B genes are linked on the same plasmid molecule yet transcripts from the early gene increase tenfold during late cleavage and early blastula stages and drop by a factor of ten after the late blastula stage while transcripts from the late gene become detectable only at the early blastula stage and increase by 25-fold during blastula stage. While minor deviations from normal occur, these results show that proper regulation of early and late H2B gene expression occurs within 600 bp of 5′ and 700 bp of 3′ flanking sequence for the early gene and 3 kb of 5′ and 100 bp of 3′ flanking sequence for the late gene. Furthermore, since different transcript accumulation patterns are observed from the two genes, the close proximity of the two promoters apparently does not effect proper transcription. The fact that the early H2B gene is removed from its environment of tandemly repeated histone gene sequences implies that this structure, per se, does not play a major role in early H2B gene regulation. The work focuses on the importance of *cis* acting regulatory elements mapping to within a few hundred bases of the CAP sites of the two genes as essential features of the early to late histone gene switch.
William H. Klein

Spatially Deranged though Temporally Correct Expression of a *Strongylocentrotus purpuratus* Actin Gene Fusion in Transgenic Embryos of a Different Sea Urchin Family

R. R. Franks, B. R. Hough-Evans, R. J. Britten, and E. H. Davidson

Genes Dev., 2, 1—12, 1988 2-9

The *Strongylocentrotus purpuratus* CyIIIa-CAT (chloramphenicol acetyl transferase) construct was used in studies directly introducing exogenous DNA into the zygote nucleus. DNA was injected into the transparent eggs of another sea urchin species, *Lytechinus variegatus,* in order to visualize the target nuclei. The embryonic expression of CAT mRNA and enzyme protein was monitored.

The CyIIIa-CAT gene was activated at the appropriate stage of development in either nuclear or cytoplasmic *L. variegatus* injection samples. However, the spatial regulation of the CyIIIa-CAT construct was abolished in *L. variegatus* host embryos. CAT mRNA accumulation no longer was confined to cells of the aboral ectoderm, as revealed by *in situ* hybridization. Copious amounts of CAT transcripts were found in skeletogenic mesenchymal cells, gut cells, and oral ectoderm. In transgenic *S. purpuratus* controls these cell types are devoid of CAT transcripts.

Regulatory sequences of the *S. purpuratus* CyIIIa actin gene mediate an ubiquitous spatial pattern of gene expression during development when introduced into *L. variegatus* eggs, contrasting with the usual aboral ectoderm-specific pattern. Temporal activation of expression, however, occurs at the appropriate stage of development in transgenic embryos. Current studies of the cytoskeletal actin genes of *L. variegatus* are intended to show whether there is an aboral ectoderm-specific member of this family.

♦The generation of transgenic animals has become the standard experimental procedure for investigating the *in vivo* temporal and spatial expression of a particular gene. In this paper, the authors use a sea urchin actin gene construct containing regulatory sequences known to properly direct both the temporal and spatial expression of the actin gene in the homologous sea urchin species. However, when introduced into a heterologous species of sea urchin, the temporal and spatial regulation become dissociated. While the actin construct is expressed at the appropriate developmental time, it is expressed in the wrong tissues. This observation has important implications for defining the actual endogenous regulatory regions of various transgenes and suggests a valuable way of exploring evolutionary alterations in the developmental regulation of specific genes. *Joel M. Schindler*

Spatially Regulated Expression of Chorion Genes During *Drosophila* Oogenesis
S. Parks and A. Spradling
Genes Dev., 1, 497—509, 1987 2-10

Construction of the *Drosophila* eggshell is a useful system for analyzing the genetic regulation of morphogenesis at a molecular level. The ultrastructure is well defined, and many of the structural genes concerned have been cloned. The authors used *in situ* hybridization to study the expression of ten RNAs transcribed from the two major *Drosophila* chorion gene clusters during oogenesis. Formation of the layered structure of the chorion by simple apposition requires precise timing of chorion gene expression.

Six abundant chorion protein mRNAs were produced through most of the follicular epithelium. Three other RNA species thought to encode minor chorion proteins accumulated in a spatial pattern related to the formation of eggshell specializations such as the dorsal appendages, the operculum, and posterior areopyles. All five chorion transcripts from an X-linked cluster appeared in a small group of dorsal follicle cells near the oocyte-nurse cell boundary. Gene-specific spatial patterns of expression then appeared in parallel with rapid centripetal and anterior follicle cell migrations.

These findings suggest that chorion genes are controlled — at least indirectly — by genes determining the spatial organization of the egg chamber in *Drosophila*. Spatial regulation by chorion genes may depend on cellular interactions that generate a coordinate system permitting the cells to determine their relative positions. The precise spatial regulation of tissue-specific gene expression probably is a common phenomenon.

♦The *Drosophila* chorion (egg shell) is synthesized at the end of oogenesis (stages 10 through 14) by a group of about 1000 follicle cells that surround the oocyte/nurse cell complex. The chorion is highly structured. It consists of a thin innermost chorionic layer followed by a thick and complex endochorion which in turn is surrounded by a thin, amorphous exochorion. The endochorion consists of a floor and a roof connected by pillars; this arrangement allows for air to circulate and provide respiration to the developing embryo. The different layers of the chorion are secreted sequentially by the overlying follicle cells in a well-defined temporal pattern. In addition to the layers, the chorion has important regional specializations. The anterior dorsal appendages and posterior areopyles are used for respiration of the embryo, the micropyle forms the opening through which sperm enters the egg at fertilization, and the operculum forms a weakened area of the chorion through which the larva hatches at the end of embryogenesis.

Of the estimated 20 structural chorion protein genes, 9 have been cloned and characterized. They are located in two clusters, one on the X and one on the third chromosome. Their temporal pattern of expression had previously been characterized. However, no information was available on their spacial pattern of expression. The asymmetry of the egg shape and the regional specializations of the chorion made it likely that some specialized chorion genes may be synthesized by spatially restricted groups of follicle cells. This hypothesis was addressed in the present paper, using *in situ* hybridization of cloned chorion probes to egg chamber sections.

The four major transcripts of the third chromosome cluster that probably synthesize the outermost layers of the chorion were expressed by almost all follicle cells. The five genes of the X chromosome cluster were expressed only in a small subset of cells early in choriogenesis, prior to extensive follicle cell migrations. The two major genes of the X cluster are then expressed in most cells. Three minor X chromosome genes, however, are expressed only in small groups of follicle cells where the eggshell specializations arise.

This study adds for the first time a new, spacial dimension to the regulation of chorion gene expression. It is now clear that the follicle cells, which earlier in oogenesis consisted of a morphologically uniform layer of cells, must respond to spacial as well as temporal clues, perhaps emanating from the underlying oocyte. Cell-cell communication must play an important role in directing groups of follicle cells to form specialized structures. When found, mutants that affect only these structures may provide clues to the mechanisms involved. *Marcelo Jacobs-Lorena*

Two-Tiered Regulation of Spatially Patterned *engrailed* Gene Expression during *Drosophila* Embryogenesis

S. DiNardo, E. Sher, J. Heemskerk-Jongens, J. A. Kassis, and P. H. O'Farrell

Nature (London), 332, 604—609, 1988 2-11

Early in generation of the *Drosophila* body pattern, the segmental subdivisions of the embryo are rapidly determined by the segment polarity genes *engrailed* and *wingless*. These genes are expressed in their localized patterns as a result of an early regulatory cascade. *Engrailed* (*en*) expression is established at the cellular blastoderm stage in single cell-wide stripes transecting the anteroposterior axis of the embryo. The expression pattern of *en* is maintained and modified in the postcellular blastoderm phase of development, by a *cis*-regulatory program distinct from that controlling expression at the cellular blastoderm stage.

The late-acting program is controlled through the activity of various segment polarity genes, at least one of which (*wingless*) is not cell autonomous. Studies utilizing an *en-lacZ* fusion gene indicated that there are distinct early and late regulatory programs controlling the expression of *en* in stripes.

It is likely that the pair-rule genes directly regulate the late program of *en* expression, because products encoded by these genes are not detected at this stage. The stage at which segment polarity gene mutations affect *en* expression coincides with the transition from early to late *en* regulatory programs.

There would seem to be considerable plasticity in developmental programming at the cellular blastoderm. The later program may represent an editing function correcting mistakes that are made in the establishment of *en* expression. It also may be that a separate early program is an evolutionary adaptation for the very rapid embryonic development of *Drosophila*. The late program would be phylogenetically conserved, while the rapid regulatory interactions leading to the early program serve as an accessory to establish rapidly spatial subdivisions.

♦As the *Drosophila* embryo cellularizes, the *engrailed* (*en*) gene is expressed in 14 single-cell-wide circular stripes distributed along its anterior-posterior axis. As the embryo develops through the germ band elongation and shortening phases, this basic pattern of *en* expression in 14 stripes is maintained. Major findings of the paper were that the establishment of *en* expression in 14 stripes is due to at least two separate promoter elements, one for even-numbered stripes and one for odd-numbered stripes, and that the maintenance of this pattern of 14 stripes is due to still another set of regulatory elements. Other experiments also suggest an intriguing model for cross regulatory interactions among segments polarity genes, including *en*.

The authors deleted approximately 3 kb from an "upstream" regulatory region of *en* and fused this modified promoter to the beta-galactosidase reporter gene. This recombinant gene was then transformed into the *Drosophila* germ line. Interestingly, early expression of the gene (beta-galactosidase activity) was restricted to, and precisely aligned with, the seven even-numbered *en* stripes (odd stripes were missing). This shows that expression in even- and odd-numbered *en* stripes is independently regulated and that the deleted region must overlap with the odd-stripe control region. Further experiments provided evidence that this control is linked to the expression of pair-rule genes. Surprisingly, in later embryos the recombinant gene is expressed in 14 stripes rather than 7. This provides evidence for a second (late) program of *en* regulation, probably involving the action of other segment polarity genes. The effects of two segment polarity mutants, *wingless* (*wg*) and *patch* (*ptc*), on *en* expression were examined for this report. The results suggested a model whereby in later embryos the expression of *en* is induced at a distance by the *wg* gene product (even though *wg* and *en* are expressed in different cells *wg* is required for an expression), while *ptc* would play a role in repressing *en* expression outside the normal *en* domain (*en* is expressed ectopically in *ptc* mutants). Consistent with the assumption that *wg* acts at a distance is the fact that *wg* mutations have been shown in genetic mosaics to be noncell autonomous. Although not all aspects of the model are firmly established, the observations reported in this paper provide new and testable ideas on possible mechanisms of action of developmentally important genes. *Marcelo Jacobs-Lorena*

Differential 5S RNA Gene Expression In Vitro

A. P. Wolffe and D. D. Brown
Cell, 51, 733—740, 1987 2-12

In the genome of *Xenopus laevis*, oocyte 5S RNA genes are 50-fold more abundant than somatic 5S RNA genes. In somatic cells, however, the minor somatic 5S RNA multigene family encodes more than 95% of the 5S RNA. Attempts have been made to reproduce this high differential gene expression using cloned 5S RNA genes *in vitro*. Studies were done utilizing an activated egg extract that exaggerates the relative transcription efficiency of cloned 5S RNA genes. Only the oocyte 5S RNA gene transcription complexes are destabilized in this extract.

Trans-acting factors in destabilized complexes are in equilibrium with free factors in this extract. Transcription, therefore, is dependent on the concentration of limiting factor in the extract. By reducing the limiting factor TFIIIA, a positive transcription factor, a cloned somatic 5S RNA gene was transcribed up to 400-fold more efficiently than a cloned oocyte 5S RNA gene.

There probably is biologic significance in the observation of an inherent difference between transcription complexes formed on the two types of 5S RNA genes. It may be that oocyte 5S RNA gene complexes are only stable *in vivo* in oocyte nuclei, while in other cell types they require a very high level of transcription factors to be activated and maintained. Alternately, the mechanisms destabilizing oocyte 5S RNA gene complexes may be necessary only to establish the somatic pattern of 5S RNA gene expression during embryogenesis, while other processes stabilize and propagate the resultant state of gene expression.

♦The key to the developmental switch from oocyte to somatic 5S RNA gene transcription is the presence in egg and oocyte cytoplasm of an activity that removes transcription complexes from oocyte but not somatic 5S RNA genes. When the germinal vesicle breaks down at maturation, this activity, associated with ribosomes, leads to the eventual inactivation of oocyte 5S genes as the supply of the primary transcription factor, TFIIIA, declines. The extent to which this switch mechanism is paradigmatic for developmental alterations in gene expression remains to be seen, but it is certainly a valuable standard for comparison. *Thomas D. Sargent*

Introns Increase Transcriptional Efficiency in Transgenic Mice
R. L. Brinster, J. M. Allen, R. R. Behringer, R. E. Gelinas, and R. D. Palmiter
Proc. Natl. Acad. Sci. U.S.A., 85, 836—840, 1988 2-13

Many mammalian genes that code for mRNA are interrupted by noncoding sequences, or introns, which may be larger than exons. Introns may function to accelerate the evolution of proteins with different properties. They also allow the differential joining of exons during splicing, and this can lead to the synthesis of variant proteins having new properties. It has been proposed that introns may be required for the efficient processing and transport of mRNA to the cytoplasm. The authors tested different genes with and without their introns and with identical 5′ and flanking regions to determine whether introns consistently influence gene expression in transgenic mice.

Four pairs of gene constructs were compared for expression of mRNA after introduction into the murine germ line through microinjection of fertilized eggs. Expression of two chimeric rat-mouse genes was assayed in fetal liver and pancreas. Two natural genes were assayed in fetal liver. In each instance, 10- to 100-fold more mRNA was produced from the intron-containing construct than from that lacking introns. Levels of mRNA paralleled the relative rates of transcription measured in isolated nuclei.

When the expression of two mouse metallothionein I gene-based constructs was studied after transection into cultured cells, little difference was noted.

Introns appear to facilitate the transcription of microinjected genes, and they may help maintain transcriptional activity during development. Introns might contain DNA sequences that are recognized at some stage of development, but are not required after transfection into established cell lines. Alternately, introns or exons may contain sequences that are important in phasing nucleosomes relative to promoter elements.

♦The ability to produce transgenic mice has been particularly useful for studying regulation of gene expression. In addition, it has been possible to redirect expression of genes to different cell types by combining the regulatory elements of one gene with the coding region of another, and this has made it possible to ask specific questions regarding gene expression and function during development. Many of the genes used for these past studies have been derived from cDNA clones. The reasons for this have been availability of clones as well as the fact that genomic clones are often too large to manipulate conveniently. These investigators present evidence indicating that cDNA clones may not yield the best results. What has been summarized in this paper are data suggesting that introns apparently increase transcriptional efficiency in transgenic mice. Four different pairs of constructs were made, each pair being identical except that one member lacked all introns. Those with introns had improved transcriptional efficiency (10- to 100-fold higher) than those without introns. Although the authors believe that the major effect of introns is on transcription rather than RNA processing, their exact role is not understood. Possibilities include important control elements that are located within the intron, or alternatively, the presence of sequences which are needed for phasing of nucleosomes relative to promoter elements. In addition to providing a practical aspect to consider when preparing gene constructs to be used in transgenic experiments, these data raise questions regarding the role of intron sequences.
Terry Magnuson

Transgenes as Probes for Active Chromosomal Domains in Mouse Development

N. D. Allen, D. G. Cran, S. C. Barton, S. Hettle, W. Reik, and M. A. Surani

Nature (London), 333, 852—855, 1988 2-14

Active chromosomal domains can influence the temporal and spatial program of gene expression. They are detected using transgenes that integrate randomly through the genome, since their expression is affected by chromosomal position. Positional effects are expected to be most apparent on transgenes lacking strong promoters, enhancers, and other modulating sequences.

The authors examined positional effects using a transgene with the weak herpes simplex virus thymidine-kinase promoter, linked with the *lacZ* indicator gene (HSV-TK-*lacZ*). Genomic DNA was injected into the male pronucleus of fertilized mouse eggs. Each transgenic fetus with detectable expression exhibited a unique *lacZ* staining pattern. The transgene was reliably transmitted to subsequent generations.

Expression of this construct appears to depend entirely on its chromosomal position. Since the transgene is transmitted to later generations, transgenes can be very useful in probing the genome for active chromosomal regions. In this way they can serve to identify endogenous genes involved in organogenesis and pattern formation.

♦The integration site of exogenous DNA introduced into a genome has a profound effect on its expression. Depending on the molecular nature of the integrated construct, this position effect can be quite variable. In this article, the authors argue that a gene construct lacking a strong promoter, enhancer, or other important regulatory sequences would be most susceptible to position effects of integration. Such a construct would be unable to influence its own expression and would be totally under the control of its integration site domain, be it active or inactive. Using such a construct, the authors demonstrate that different transgenic lines have different developmental patterns of expression. They argue that these different patterns reflect when the chromosomal integration sites of the exogenous DNA are in active configurations. They suggest that these transgenes can now be used as probes to explore active chromosomal domains during development. *Joel M. Schindler*

The Leucine Zipper: A Hypothetical Structure Common to a New Class of DNA Binding Proteins

W. H. Landschulz, P. F. Johnson, and S. L. McKnight

Science, 240, 1759—1764, 1988 2-15

There is much reason to expect new structural motifs for DNA-binding proteins. The authors discovered an amino acid sequence motif common to five DNA-binding proteins, three nuclear-transforming proteins, and two transcriptional regulatory proteins. It consists of a periodic repetition of leucine residues. The leucines probably extend from a long α helix, with the leucine side chains of one helix interdigitating with those of a matching helix from a second polypeptide to form a stable noncovalent linkage.

A 30-amino acid segment of C/EBP, an enhancer-binding protein, shares considerable sequence similarity with a segment of the cellular Myc transforming protein. A display of the respective amino acid sequences on an idealized α helix showed a repetition of leucine residues at every seventh position over eight helical turns. The same periodic array of four or more leucines was observed in the sequences of the Fos and Jun transforming proteins, as well as the yeast gene regulatory protein GCN4.

The paired helices of this class of proteins may have an important role in arranging the contact surface for sequence-specific interaction with DNA. The "leucine zipper" facilitates dimerization. The leucine side chain has two methyl groups extending from its single γ carbon and none appended to the β carbon. The side chain, therefore, is long, symmetrical, and bulky at its tip, allowing the leucine residues from one α helix to interdigitate with those of another to form the molecular "zipper".

♦The interactions between structural and regulatory proteins and DNA remodel the genome to convey the information necessary for mitosis, meiosis, and for the temporal and tissue specific programs of development of an organism. The molecular details of these fundamental interactions are only recently being elucidated. Two protein structural motifs are known to be important in binding to DNA: the helix-turn-helix, established for the gene activator and repressor proteins of bacteria, and the zinc-coordinating repeating unit of two closely spaced cysteines followed by two histidines (zinc finger), established for a number of regulatory proteins in eukaryotic systems. Landschulz et al. describe a third motif, the leucine zipper. Their zipper model is based on a sequence comparison of five DNA binding proteins, two of which regulate transcription (CCAAT or enhancer binding protein, C/EBP, and GCN4) and three of which transform cells (myc, fos, and jun). Essential to the leucine zipper is the periodic repetition of leucines, at least four leucines spaced seven residues apart, in an α-helix of a minimum of six helical turns. Protein domains containing five leucine repeats would span eight helical turns. The predicted helices are amphipathic, although the degree of hydrophobicity of the nonpolar face of the helix is not remarkable. Based on the fact that the heptad repeat is the quintessential feature of proteins which adopt a coiled-coil structure, the authors propose that the leucine heptad repeat comprises a dimerization site in these DNA binding proteins. The leucine residues of two chains are proposed to interdigitate, leaving the hydrophilic faces of the α-helix stabilized and exposed for presentation to the DNA. Structural arguments anticipate an antiparallel orientation of the dimeric subunits. The leucine zipper model cannot itself explain the DNA sequence specificity of interaction of the dimer with the DNA. It is suggested that the amino acid sequence abutting the leucine zipper confers DNA binding specificity. This region is highly basic in C/EBP and fos. The leucine zipper model is elegant in its simplicity, and it is guaranteed to stimulate extensive experimental testing. *Judith A. K. Harmony*

Congenital Immunodeficiency with a Regulatory Defect in MHC Class II Gene Expression Lacks a Specific HLA-DR Promoter Binding Protein, RF-X

W. Reith, S. Satola, C. Herrero Sanchez, I. Amaldi, B. Lisowska-Grospierre, C. Griscelli, M. R. Hadam, and B. Mach

Cell, 53, 897—906, 1988 2-16

Expression of major histocompatibility complex (MHC) class II genes is tightly regulated. In congenital severe combined immunodeficiency (SCID), a regulatory defect precludes the expression of HLA class II genes. B lymphocyte cell lines from these patients provide a means of identifying regulatory proteins that bind to class II gene promoters.

The authors identified three proteins binding to specific segments of the HLA-DRA promoter. Two of them interact to form the predominant DNA-protein complex. One of these proteins, an X box-binding protein (RF-X), is absent in cells from class II-deficient patients with SCID.

It is proposed that RF-X is a regulatory factor required for constitutive class II gene transcription in B lymphocytes. This is the first example of a congenital disorder involving a defined molecular defect in a regulatory DNA-binding protein. SCID patients from different regions and ethnic groups have exhibited the same defect in protein binding to class II promoters. These findings could relate to the pathogenesis of certain autoimmune disorders associated with aberrant expression of class II antigens.

♦Class II deficient severe combined immunodeficiency (SCID), a congenital disorder which is characterized by the absence of expression of class II major histocompatibility (MHC) glycoproteins in all tissues, is the first hereditary disease caused by a defect in a transcription factor to be identified. The tissue distribution and developmental timing of class II MHC gene expression is tightly regulated. Normally restricted to B lymphocytes, activated T lymphocytes, monocyte-macrophages, and dendritic cells, class II MHC molecules can be induced on other cell types by γ-interferon (IFN-γ). Reith et al. have examined the promoter region of the HLA-DRA class II gene in B cell lines established from normal donors and from patients with class II SCID. The promoter region contains conserved DNA sequence domains, including a TATA box important for accurate initiation of transcription, an octamer (ATTTGCAT) sequence similar to that in immunoglobulin light and heavy chain genes but absent in other class II genes, and X and Y boxes strongly conserved in all human and murine class II genes. The function of the X and Y boxes has not been defined unequivocally, although the evidence implicates their importance as *cis*-acting elements involved in the regulation of class II MHC gene expression. Using the strategies of gel retardation assays, DNAase I footprinting and methylation interference, the investigators identified three putative regulatory proteins which bind to the HLA-DRA promotor region. Two of these, RF-X and RF-Y, appear to interact with each other when they bind to the X and Y boxes, respectively. RF-X is missing or defective in cell lines derived from class II SCID patients. Furthermore, since class II genes are not induced by IFN-γ in class II-negative SCID fibroblasts, RF-X seems to be important in constitutive and induced expression of class II genes. It is now imperative to define the molecular basis of the defect. *Judith A. K. Harmony*

Human Gene Expression First Occurs Between the Four- and Eight-Cell Stages of Preimplantation Development
P. Braude, V. Bolton, and S. Moore
Nature (London), 332, 459—461, 1988 2-17

A developmental dependence on expression of the embryonic genome is detected at the mid-2-cell stage in the mouse; the 4-cell stage in the pig; and the 8-cell stage in the sheep. The timing of activation of the human embryonic genome is relevant to *in vitro* fertilization and embryo transfer, as well as to the development of methods for preimplantation diagnosis of inherited genetic disorders.

The authors studied changes in the pattern of polypeptide synthesis in the preimplantation stages of human development. Single human oocyte and cleaving preembryo polypeptides were labeled with ^{35}S-methionine. Both losses and gains of bands were observed between the 4-cell stage and morula. A transcriptional inhibitor, alpha amanitin, precluded development beyond the 4-cell stage. When 4-cell preembryos were exposed to inhibitor, either development was arrested or only one further complete round of cleavage divisions was completed.

The first two cell cycles of human embryogenesis appear to be regulated at a posttranscriptional level, utilizing maternally inherited information. The embryonic genome is activated between the 4- and 8-cell stages. If the products of a defective gene are measured for preimplantation diagnosis, the product identified must be encoded by the embryonic genome. Preimplantation diagnosis should utilize blastomeres or tissue from preembryos after this developmental stage.

♦Little biochemical or molecular work has been done with the human embryo, therefore, any published reports on this subject would add significant information. This report is particularly important because it addresses the question of when the human embryonic genome is activated. It is known that in mouse and sheep this event takes place at the mid-two- and eight-cell stages, respectively. This report demonstrates that major qualitative changes in protein synthesis occurs between the four- and eight-cell stage of human preimplantation development and that some of these changes are sensitive to inhibitors of transcription. In conjunction with this observation, it appears that development beyond the four- to eight-cell stage is also transcription dependent. Thus, the data presented are consistent with the first two cell cycles of human embryogenesis being regulated at the post-transcriptional level, and with activation of the embryonic genome occurring between the four- and eight-cell stages. A major problem with studies dealing with human embryos is that sample sizes are limited due to the difficulty in obtaining material. Thus, results must be interpreted with some caution until repeated by others. *Terry Magnuson*

Developmental Cell Biology 3

INTRODUCTION

Development could ultimately be reduced to an issue of cell biology. While it is true, as we have already seen, that genetics and gene expression can and do influence development, it is the products of those genes that confer function and ultimately impact on cell behavior. Thus, cell behavior is the essence of development.

The relationship between cell growth and cell differentiation has been and remains a central question in developmental biology. While growth and differentiation are mutually exclusive in some organisms (e.g., slime mold), the relationship is far more complex in most others.

The cell cycle and cell cycle-related events influence normal development. So do certain cellular proteins which comprise part of the cellular architecture and which must be regulated to allow normal division to occur.

One of the most exciting areas of current interest is the observation that mammalian growth factors, which directly influence the extent of mammalian cell proliferation, have homologous "relatives" in other organisms. These homologous peptides seem to play important roles in the normal development of their respective organisms. Thus, molecules known to be important for cell proliferation seem to also be involved in the programming of normal development. The fact that such molecules have been conserved over a fairly long evolutionary period of time suggests that they play a key role in regulating one or more different cellular functions.

The papers included in this chapter directly address the role of the cell cycle in development, discuss several cellular factors (structural or otherwise) that seem to be important, and present a survey of different growth factor-like molecules and the exciting way in which they influence development.

Cell-Autonomous Determination of Cell-Type Choice in *Dictyostelium* Development by Cell-Cycle Phase
R. H. Gomer and R. A. Firtel
Science, 237, 758—762, 1987 3-1

The fate of individual cells of the eukaryote *Dictyostelium discoideum* was studied in a system allowing the cells to differentiate in the absence of aggregation. The generation of two cell types has been ascribed to a cell's position within the aggregate and to other, nonpositional factors. Prestalk and prespore cells initially are found intermixed within the developing aggregate, suggesting that nonpositional factors are at least partly responsible.

The tendency of single amoebas to differentiate into prespore or prestalk cells occurred by a cell-autonomous mechanism dependent on the position of the cell in the cell cycle at the initiation of development. Cells dividing between about 1 1/2 h before and 40 min after differentiation-inducing starvation become prestalk cells. Cells dividing at other times became prespore cells.

The signal determining the initial ratio of prestalk and prespore cells in *D. discoideum* may depend on cell cycle-related events, through a cell-autonomous mechanism. The molecular mechanism of this process remains to be learned. It could involve a substance synthesized at one phase of the cell cycle; the stability of the substance could modulate the ratio of cell types. Other possible mechanisms could involve cell-cell contact.

Cell Cycle Phase in *Dictyostelium discoideum* is Correlated with the Expression of Cyclic AMP Production, Detection, and Degradation: Involvement of Cyclic AMP Signaling in Cell Sorting

M. Wang, R. J. Aerts, W. Spek, and P. Schaap

Dev. Biol., 125, 410—416, 1988 3-2

Cell cycle phases in *Dictyostelium* correlate with a preference for spore or stalk differentiation. Cells that begin development early in the cell cycle (E cells) tend strongly to sort to the prestalk region of the slug, while late cell-cycle cells (L cells) sort to the prespore region. The authors sought insight into the cause of differences in sorting behavior between E and L cells by studying the expression of the cAMP chemotactic system during development. It was thought that differences in the production, detection, or degradation of chemotactic signals might explain sorting behavior.

E cells exhibited cAMP-binding activity, cell surface cAMP-phosphodiesterase activity, and an ability to relay cAMP signals earlier and to higher levels than L cells. Cell cycle-related variations in expression of cAMP-binding activity persisted in a mixed population of E and L cells after a period of autonomous cAMP oscillations.

These findings suggest that E cells are prestalk sorters because they respond most effectively in chemotaxis and signal relay processes.

♦It was demonstrated that *Dictyostelium* amoebae differentiate by a cell-autonomous mechanism into either prespore or prestalk cells. Cells are videotaped while growing at low density. Examination of the video recordings indicates that cells that have divided 1.5 h before and 40 min after starvation are induced to become prestalk cells. In contrast, cells that were stained with the antiprespore antibody last divided at other times. Growing cells were synchronized in two ways. Cells which were dividing or had just divided at the onset of development differentiated into predominantly prestalk cells. Nondividing cells gave rise to prespore cells. Thus, these experiments are consistent with the low density experiments. The data suggest that cell cycle-related events determine the pathway of differentiation and the cells then use other mechanisms, such as differential cohesion or motility to sort out into the proper pattern in the migrating slug.

In a related report by Wang et al. they suggest that the E cells (early cell cycle) become prestalk because they initiate the aggregation centers and respond earliest to the chemotactic signal. *Stephen Alexander*

F-Actin Binds to the Cytoplasmic Surface of Ponticulin, a 17-kD Integral Glycoprotein from *Dictyostelium discoideum* Plasma Membranes

L. J. Westehube and E. J. Luna
J. Cell. Biol., 105, 1741—1751, 1987 3-3

Actin is a major component of the submembranous network of the plasma membrane. F-actin affinity chromatography and immunologic methods are used to detect actin-binding proteins in purified plasma membranes from *Dictyostelium discoideum.* A 17-kDa integral membrane glycoprotein, gp17, has been found to be responsible for much of the actin-binding activity of plasma membranes.

The actin-binding activity of gp17 is identical to that of intact plasma membranes. Monovalent antibody fragments directed against gp17 inhibit actin-membrane binding nearly completely in sedimentation assays. Fab directed against cell-surface determinants inhibited binding by 10% or less. The actin-binding site of gp17 appeared to be located on the cytoplasmic surface of the membrane. The glycoprotein constitutes from 0.4 to 1% of total membrane protein. Extracellular determinants of gp17 are evident in addition to the cytoplasmic localization of the actin-binding site. Surface labeling of gp17 by sulfo-*N*-hydroxy-succinimido-biotin, which does not penetrate the cell membrane, was observed. The glycoprotein is glycosylated.

Glycoprotein 17 is a major actin-binding protein that serves to connect the plasma membrane to the microfilament network. It is termed "ponticulin" — from the Latin for small bridge. Through binding to membrane proteins and by providing an attachment site for actin, ponticulin may transduce signals from membrane receptors to the contractile elements of the cytoskeleton.

♦A 17-kD glycoprotein (gp 17) which specifically binds F-actin has been isolated from the plasma membranes of *Dictyostelium discoideum*. This material has F-actin binding properties that are identical to that of the intact membranes: (1) it is resists extraction with 0.1 *N* NaOH, 1 m*M* DTT; (2) it is sensitive to salt concentration but stable to pH; and (3) it is sensitive to proteolysis, heat denaturation, and treatment with DTT and *N*-ethylmaleimide. Cell surface labeling shows that gp 17 is also displayed on the outer surface of the cell. Thus, the gp 17 molecule appears to be transmembrane in nature. The gp 17 molecule is glycosylated since it is also recognized by labeled conA on western blots. The gp 17 appears to be a major actin binding protein that may be involved in linking the cell surface to the underlying microfilament network. The protein has been named "ponticulin". *Stephen Alexander*

Molecular Cloning and Characterization of the mRNA for Cyclin from Sea Urchin Eggs

J. Pines and T. Hunt
EMBO J., 6, 2987—2995, 1987 3-4

Cyclin is a protein of fertilized eggs and oocytes, observed during cleavage and meiotic maturation, respectively. It may be an essential component for cells to enter mitosis, and its destruction may be necessary to exit from mitosis. The authors isolated a cDNA clone encoding sea urchin cyclin and determined its sequence.

The cDNA clone encoding cyclin contains a single open reading frame of 409 amino acids, exhibiting homology with clam cyclins. RNA transcribed from the sequence was efficiently translated in reticulocyte lysates, yielding full-length cyclin. Injection of the synthetic mRNA in nanogram amounts into *Xenopus* oocytes caused them to mature more rapidly when on progesterone treatment. Sea urchin cyclin underwent two post-translational modifications during maturation in *Xenopus* oocytes. The first was at about the time that maturation became resistant to cycloheximide, and the second involved destruction of cyclin at about the time of appearance of the white spot.

Cyclin is the only sea urchin maternal mRNA capable of inducing frog oocyte maturation in the microinjection assay. Cyclin may act by activating maturation promoting factor (MPF) through inactivating a protein phosphatase or some negative regulator of MPF activation. When cyclin is destroyed, the protein phosphatase is again active and MPF will be converted back to pro-MPF.

♦Xyclin was first discovered in sea urchin and clam eggs by Hunt and coworkers (Evans et al., *Cell*, 33, 389—396, 1983), and has since been investigated extensively in clams by Ruderman and colleagues. In clams, two cyclins, A and B, are present. These proteins were so named because they are synthesized and destroyed at certain points in the cell division cycle. The cyclin A of clams has been shown to induce meiotic maturation in *Xenopus* oocytes and it is believed that this protein is part of a cascade that causes cells to enter into meiosis or mitosis. Though both cyclin genes have been cloned from clam, the sea urchin cyclin gene has been difficult to clone. This article describes the isolation and characterization of a cyclin cDNA clone and its corresponding mRNA from *A. punctulata* and also extends the observations on the function of this protein. An interesting feature of the cyclins is that their sequences do not appear to be highly conserved. Except for a region in the center of the molecule, the sea urchin sequence shows only a weak match with clam cyclin A, although both apparently have the same function. Besides the sequence comparisons, the noteworthy results presented here are that sea urchin cyclin can cause meiotic maturation in frog oocytes; cyclin is destroyed in the oocytes at a specific time after the onset of maturation, when the meiotic metaphase spindle forms; and sea urchin maternal mRNA probably contains no other sequences capable of promoting oocyte maturation. Taking into account what is known about MPF and its activation, the authors favor the idea that cyclin acts by inactivating a negative regulator of MPF activation. *William H. Klein*

A Sea Urchin Gene Encodes a Polypeptide Homologous to Epidermal Growth Factor

D. A. Hursh, M. E. Andrews, and R. R. Raff

Science, 237, 1487—1490, 1987 3-5

Epidermal growth factor (EGF) is a member of a family of proteins that have functions relating to cellular differentiation, proliferation, and transformation. This report describes the isolation of an EGF-related gene from sea urchins.

A screen for cell lineage-specific genes from *Strongylocentrotus purpuratus* yielded a 1.5 kb cDNA clone, which contained a 1447 nucleotide open reading frame and hybridized to a 3.0- and a 4.0-kb RNA. A search of the Protein Information Resource data base indicated strong homology between the open reading frame sequence and that of epidermal growth factor-like proteins. Both contain a characteristic repeat with the pattern, CX4CX5CX8CXCX8CX6, (C is a cysteine residue and X any residue). This repeat is found nine times in the sea urchin gene, uEGF-1. The two transcripts recognized by uEGF-1 are highest in unfertilzed eggs, decline during early cleavage, increase at blastula, and then decline after gastrula.

The similarity of uEGF-1 to other members of the EGF protein family, as well as its unusual pattern of expression, suggest that uEGF-1 may play an important role in embryogenesis. The existence of this highly conserved sequence in echinoderms indicates that this conserved peptide domain predates the radiation of coelomate animals and has been retained in diverse evolutionary lineages.

♦These investigators have found a transcript sequence in the sea urchin embryo which contains similarity to the epidermal growth factor (EGF) family of proteins. This is the first evidence for presence of a growth factor in the sea urchin and is important for several reasons. First, these results extend the list of invertebrates containing EGF family members. What was previously thought to be exclusive to mammals now includes the *notch* gene of *Drosophila* (Wharton et al., *Cell,* 43, 567, 1985), the *lin 12* gene of *Caenorhabditis elegans* (Greenwald, *Cell,* 43, 583, 1985), and the sea urchin sequence reported here. The EGF family now appears to be widely used in metazoan development. A second important feature of these findings is the developmental distribution of the EGF transcript. The message appears to accumulate coincident with differentiation of the ectodermal germ layer. Although the sequence is present in greatest abundance in the egg, its spatial and temporal distribution is consistent with a role in the development and differentiation of the tissue in the sea urchin analagous to an epidermis. Sequence analysis of the sea urchin EGF clone indicates it contains a feature of all EGF family members: it contains the characteristic cysteine distribution of $CX_4CX_5CX_8CXCX_8CX_6$, found in nine repeat units in this species. A third intriguing aspect of these findings is speculative. Does this result imply that other growth factors also may be present during development and differentiation in the sea urchin or other invertebrates? *Gary M. Wessel*

The Embryonic Expression of the *Notch* Locus of *Drosophila melanogaster* and the Implications of Point Mutations in the Extracellular EGF-Like Domain of the Predicted Protein

D. A. Hartley, T. Xu, and S. Artavanis-Tsakonas

EMBO J., 6, 3407—3417, 1987

3-6

The *Drosophila melanogaster Notch* locus is a zygotic neurogenic gene, which is required for correct segregation of neural and epidermal lineages during embryogenesis. The expression and regulation of the *Notch* gene was examined at the molecular level.

In situ hybridization with two probes, one which contains the EGF-like repeats and one which does not, was used to investigate wild-type *Notch* expression. *Notch* was expressed throughout the early embryo. After segregation of neuroblasts from ectoderm, *Notch* continued to be expressed in both neural and epidermal precursors. To determine whether *Enhancer of split [E(spl)]* regulates *Notch* transcription, *Notch* probes were hybridized to loss of function *E(spl)* mutants. The distribution of *Notch* mRNA in the mutants was similar to wild type. Since transcriptional control did not appear to account for the interaction between these two genes, a *split* mutant gene was sequenced. The only change in the sequence which altered the predicted amino acid sequence was a change in amino acid 578, from isoleucine to threonine. The mutation occurs in the EGF-like extracellular region of this protein. An *Abruptex* mutation was also sequenced and determined to be a point mutation in the extracellular EGF-like region of the protein as well.

The unexpectedly widespread expression of *Notch* during embryogenesis indicates that *Notch* may have a larger role in development than has been predicted and that its function is context dependent. The interaction of *E(spl)* with *Notch* does not appear to occur at the level of transcription. Their gene products may interact directly, as a point mutation in the extracellular portion of the *Notch* protein disrupts this interaction. Antibodies to various regions of the *Notch* protein are being used to analyze its expression and function further.

The Neurogenic Gene Delta of *Drosophila melanogaster* is Expressed in Neurogenic Territories and Encodes a Putative Transmembrane Protein with EGF-Like Repeats

H. Vassin, K. A. Bremer, E. Knust, and J. A. Campos-Ortega
EMBO J., 6, 3431—3440, 1987 3-7

One of the *Drosophila melanogaster* neurogenic genes is Delta (*Dl*). This paper describes the cloning and characterization of this locus.

Over 180 kb of DNA surrounding the Delta locus was cloned. The region containing *Dl* was narrowed down to 25 kb by mapping mutations. By Northern blot analysis, two developmentally regulated mRNAs of 5.4 and 4.6 kb, associated with this region, were detected. The complete sequence of a 4.73-kb cDNA clone homologous to this region was determined. The longest open reading frame yielded a protein of 880 amino acids, with intracellular, transmembrane, and extracellular domains. The extracellular domain contained nine EGF-like repeats. *In situ* hybridization with the cDNA clone indicated that *Dl* was initially detected in all cells with neurogenic potential. *Dl* transcription gradually became restricted to cells that adopted the neurogenic fate.

The cellular location of *Dl* expression is exactly what would be predicted for a gene that regulates the decision of an ectodermal cell to take on a neural or an epidermal fate.

♦These two papers provide interesting similarities and contrasts in the study of two "neurogenic" loci in *Drosophila. Notch (N)* and *Delta (Dl)* are two of five zygotic "neurogenic" loci that when mutated yield a very similar embryonic phenotype: the hyperplasia of neuroblasts at the expense of epidermal cells. The similarity of the phenotypes suggested that these genes act in concert to direct developing embryonic cells along the mutually exclusive neuronal or epidermal pathways. The genetics of the *N* locus are complex. The large number of mutations in the locus include the recessive muatation *split (spl)* that causes the scarring or roughening of the eye, and the dominant mutations of the *Abruptex (Ax)* group that exhibit gapped veins and bristle abnormalities. It is possible that part of the genetic complexity is due to the formation of homomeric and heteromeric protein complexes that are required for function. The *N* locus has been cloned and sequenced. The presumptive *N* protein has several unique features. A hydrophobic sequence typical of a transmembrane domain is found in the center of the protein, the presumptive extracellular portion of the protein is composed of 36 tandem 40-amino-acid repeats with homology to mammalian epidermal growth factor (EGF), and on the cytoplasmic side are sequences reminiscent of a nucleotide binding site. Hartley et al. have examined by *in situ* hybridization the distribution of *N* transcripts in the embryo. Surprisingly, they found that the transcripts are not limited to neurogenic or epidermal cells but that they are distributed throughout the embryo. These results are hard to interpret but imply that *N* plays a more general role in embryonic differentiation. In a separate set of experiments, Hartley et al. sequenced the *N* gene from *spl* and *Ax* mutants. Interestingly, they found that each mutation corresponds to a single amino acid change in a different EGF repeat. In case of *spl* this may identify the site of interaction with the protein of a separate locus, *Enhancer of split,* that genetically interacts with *spl.*

The paper by Vassin et al. reports the characterization of *Dl.* Interestingly, *Dl* codes for a protein with a putative transmembrane domain containing nine tandem arrays of EGF-like repeats in the presumed extracellular domain. In contrast with *N, in situ* hybridization reveals that expression of *Dl* in early embryos is almost exclusively confined to the presumptive neurogenic regions, and in older embryos accumulation of *Dl* transcripts is restricted to cells that have adopted the neural fate.

Certainly the characteristics of these neurogenic genes raised some intriguing and unexpected questions. The active research in this area promises exciting findings in a not too distant future, especially in what concerns the now poorly understood role of the EGF-like protein domains. *Marcelo Jacobs-Lorena*

Role of Nuclear Material in the Early Cell Cycle of Xenopus Embryos

M. C. Dabauvalle, M. Doree, R. Bravo, and E. Karsenti

Cell, 52, 525—533, 1988 3-8

Although activated or fertilized *Xenopus* eggs that have been enucleated fail to divide, they undergo periodic surface contraction waves (SCWs) that are synchronous with the cleavage cycle of eggs which have not been enucleated. To determine if nuclear components are part of the cell cycle oscillator, maturation promoting factor (MPF) content and phosphorylation were examined in eggs from which the germinal vesicle had been removed.

Oocytes were enucleated, treated with progesterone for 1 h, incubated overnight at 19° C, and then activated by pricking. SCWs were monitored by time-lapse video microscopy. In the absence of germinal vesicle material, SCWs were temporarily suppressed. Microhomogenates were prepared and injected into recipient oocytes to monitor MPF content. MPF activity increased and decreased with the same kinetics in nucleated and enucleated eggs. High MPF activity has been correlated with high levels of protein phosphorylation. Protein phosphorylation occurred with the same kinetics in nucleated and enucleated eggs. Two proteins, M116 and M46, which are known to become phosphorylated during metaphase, were not phosphorylated in enucleated eggs and are probably located in the germinal vesicle. Histone kinase activity oscillated with the same kinetics in nucleated and enucleated eggs.

These results demonstrate that the regulation of the oscillation of MPF and kinase activities occurs in the absence of the germinal vesicle. Therefore, the essential cell cycle oscillator appears to be located exclusively in the cytoplasm. The temporal regulation of MPF and kinase activity is likely to be this cytoplasmic cell cycle oscillator.

♦The timing and frequency of the cell cycle has important developmental implications. During *Xenopus* development, the cell cycle goes through a transition from a simple cycle composed of only an M and an S phase to a more complex cycle that includes the more typical G1 and G2 phases. This transition occurs at the mid-blastula stage, a time just prior to the onset of several important developmental events, including transcription and cell motility. A major question regarding this transition has been the role of nuclear vs. cytoplasmic components in the process. The work reported in this paper addresses this question. The authors use activated, enucleated eggs to monitor several cellular events that normally occur following egg activation. They show that both the temporal control of the maturation promoting factor (MPF) and kinase activities occur normally in the absence of nuclear material, but surface contraction waves do not. This suggests that MPF and kinase activities are important features of the cell cycle oscillator, but calls into question the role of nuclear components in regulating the timing of the mid-blastula cell cycle transition. *Joel M. Schindler*

The Organization of Mesodermal Pattern in *Xenopus laevis*: Experiments Using a *Xenopus* Mesoderm-Inducing Factor

J. Cooke, J. C. Smith, E. J. Smith, and M. Yaqoob

Development, 101, 893—908, 1987 3-9

Mesoderm-inducing factor (MIF), which is secreted by the *Xenopus* XTC line, leads to the differentiation from competent ectoderm of mesodermal structures characterizing the dorsal parts of the body. Several different types of experiments were performed to determine if MIF is a natural initiator of mesoderm formation in *Xenopus.*

Freshly excised blastula animal caps were exposed to XTC-conditioned medium for varying lengths of time. Explants whose inner membranes were exposed for more than 10 min demonstrated differentiation to mesoderm. Partially purified MIF was microinjected into blastocoels. Irrespective of dose and timing, the entire blastocoel roof appeared to form mesoderm. There was some recovery of normal gastrulation, depending upon the dose injected. Either MIF or a control dialysate was injected into sibling embryos, either intracellularly, into the animal hemisphere at the 1- to 4-cell stage or into the blastocoel at the 128-cell stage. After intrablastocoelic injection, ectopic mesoderm was formed. However, if this same dose was injected into the animal hemisphere or the 2- to 4-cell embryo, it had no effect. Therefore, microinjected MIF was not effective or secreted. However, if poly A+ mRNA from XTC cells was injected, mesoderm was induced. It appears that newly synthesized MIF was capable of being secreted from blastomeres and was effective once outside the cell. Animal pole ectoderm was dissected from embryos that had received MIF blastocoel injections. These tissues were grafted to the ventral marginal zones of stage 9 embryos. In the majority of embryos, secondary caudal axes developed. This suggests that MIF-treated animal pole ectoderm acts as a Spemann-Mangold organizer.

The effects of MIF on competent animal pole tissue are consistent with a role for MIF as the natural *Xenopus* mesoderm-inducing factor. However, other factors are probably involved in the establishment of the mesodermal pattern.

♦There are four important results in this paper dealing with partially purified preparations of the mesoderm inducer Jim Smith discovered in medium conditioned by *Xenopus* Tissue Culture (XTC) cells (J. C. Smith, *Development,* 99, 3—14, 1987). First, exposure of competent animal pole cells to XTC factor for as little as 15 min is sufficient to respecify the ectoderm as mesoderm. This is interesting because the time required for this kind of specification to be brought about by recombining animal and vegetal explants is much longer, on the order of 1 to 2 h (Gurdon et al., *Cell,* 47, 913—922, 1985). Second, a piece of ectoderm exposed to XTC factor can act as a Spemann-Mangold organizer when grafted into a competent host, which is further support for the idea that XTC factor is a potent dorsalizing agent as well as a mesoderm inducer. Third, injection of XTC factor into cells has no inductive effect, indicating that a surface receptor interaction is required. Fourth, injection of mRNA from the XTC cell line causes the recipient cells and their neighbors to turn into mesoderm. This is a good control for the factor injection experiment, and it also means that injection of unfractionated mRNA can be used as an assay for the presence of mRNAs encoding inducers, which could be useful in cloning strategies. Woodland and Jones (*Development,* 101 925—930, 1987) have reported similar results with XTC mRNA. *Thomas D. Sargent*

Synergistic Induction of Mesoderm by FGF and TGF-b and the Identification of an mRNA Coding for FGF in the Early Xenopus Embryo

D. Kimelman and M. Kirschner

Cell, 51, 869—877, 1987 3-10

It has been found that mammalian growth factors, such as fibroblast growth factor (FGF), can induce the formation of mesoderm from *Xenopus* animal cap cells. However, it is not known whether such factors are the actual embryonic inducers. In this paper, the presence and role of growth factor-like molecules was investigated in *Xenopus* embryos.

Animal caps were incubated in the presence and absence of various mammalian growth factors and the induction of a mesoderm-specific mRNA (cardiac actin) was assessed by RNAse protection assays. Although transforming growth factor (TGF-beta) and epidermal growth factor (EGF) had no effect, basic bovine FGF induced cardiac actin expression when added before gastrulation and assayed at the neurula stage. Cardiac actin expression was induced by FGF from midblastula to the onset of gastrulation and occurred at stage 14, which is the normal activation time. The maximum response to FGF occurred at 25 ng/ml, but was only 5 to 10% of that seen in intact embryos. To examine whether more than one factor was required for complete mesoderm induction, growth factor combinations were tested. Addition of TGF-beta, 7 ng/ml, increased cardiac actin expression tenfold over the level induced by FGF alone.

To determine whether bovine FGF was related to the natural *Xenopus* inducer, a cDNA for bovine basic FGF was used to isolate a clone from a *Xenopus* genomic library. This clone was used to screen a *Xenopus* oocyte cDNA library. The resulting clones were sequenced. An open reading frame encoding a peptide that was 89% homologous to the C terminus of FGF was found. Therefore, a gene encoding a protein that is related to basic bovine FGF is expressed during *Xenopus* oogenesis. Blots of polyA+ RNA were used to determine that FGF-like mRNA is present throughout *Xenopus* early development.

These results indicate that a FGF-like protein is part of the natural mesoderm inducer in *Xenopus*. The *Xenopus* egg provides a method to test directly the function of growth factors during development.

♦This work continues the small revolution begun last year with the discovery that mammalian growth factors can act as mesodermal inducers in *Xenopus*. In this paper evidence is presented that supports the notion that there are two distinct mesoderm induction processes, one, mediated by fibroblast growth factor (FGF), tends to elicit ventral mesoderm, and another, mediated by transforming growth factor beta (TGFβ), results in dorsal and lateral mesoderm, such as notochord and somites. The evidence is especially strong for a role of FGF, as mRNA encoding this protein is found in the egg and embryo, and the *in vitro* inducing effects are reproduced with cloned FGF manufactured in bacteria, eliminating the possibility of contaminating factors. Nevertheless, there is some disagreement as to which factors are important; see Rosa et al., 1988. *Thomas D. Sargent*

A Maternal mRNA Localized to the Vegetal Hemisphere in Xenopus Eggs Codes for a Growth Factor Related to TGF-β

D. L. Weeks and D. A. Melton

Cell, 51, 861—867, 1987

3-11

The *Xenopus* protein Vg1 is encoded by maternal mRNA and is localized to presumptive endoderm. Therefore, it is a candidate for involvement in developmental commitment. This paper describes the sequencing and characterization of Vg1.

A previously identified cDNA was used to screen a library to obtain a full-length clone. The sequence of Vg1 is shown in Figure 1 (*Cell*, 51, 862, 1987). The deduced amino acid sequence was used to search the National Biomedical Research Foundation protein sequence data bank. Homology to the TGF-beta protein family was found (see Figure 3-11).

It has recently been shown that TGF-beta can participate in mesoderm induction in the *Xenopus* embryo. Together with this data, these results suggest that Vg1 is involved in the specification of mesoderm during *Xenopus* development.

♦In view of the recently published indications (Kimelman and Kirschner, 1987; Rosa et al., 1988) that TGFβ or factors related to it can induce dorsal mesoderm, it is quite interesting that Weeks and Melton have identified the single vegetal-specific mRNA they were able to clone as a member of the TGFβ family. There are no data yet regarding the actual function of the protein encoded by Vg1, but it would be astounding if this growth-factor-like polypeptide were not involved in mesoderm induction in some important way. *Thomas D. Sargent*

```
                                                                            *
Vg1 protein         Arg Pro Arg Arg Lys Arg Ser Tyr Ser Lys Leu Pro Phe Thr Ala Ser Asn Ile Cys Lys
TGF-β protein       Arg His Arg Arg Ala Leu Asp Thr Asn Tyr Cys Phe Ser Ser Thr Glu Lys Asn Cys Cys
β-Inhibin protein   His Pro His Arg Arg Arg Arg Arg Gly Leu Glu Cys Asp Gly Lys Val Asn Ile Cys Cys
DPP protein         Arg Asn Lys Arg His Ala Arg Arg Pro Thr Arg Arg Lys Asn His Asp Asp Thr Cys Arg

                    Lys Arg His Leu Tyr Val Glu Phe Lys     Asp Val Gly Trp Gln Asn Trp Val Ile Ala
                    Val Arg Gln Leu Tyr Ile Asp Phe Arg Lys Asp Leu Gly Trp Lys     Trp Ile His Glu
                    Lys Lys Gln Phe Phe Val Ser Phe     Lys Asp Ile Gly Trp Asn Asp Trp Ile Ile Ala
                    Arg His Ser Leu Tyr Val Asp Phe Ser     Asp Val Gly Trp Asp Asp Trp Ile Val Ala

                                                     *               *
                    Pro Gln Gly Tyr Met Ala Asn Tyr Cys Tyr Gly Glu Cys Pro Tyr Pro Leu Thr Glu Ile
                    Pro Lys Gly Tyr His Ala Asn Phe Cys Leu Gly Pro Cys Pro Tyr Ile Trp Ser     Leu
                    Pro Ser Gly Tyr His Ala Asn Tyr Cys Glu Gly Glu Cys Pro Ser His Ile Ala Gly Thr
                    Pro Leu Gly Tyr Asp Ala Tyr Tyr Cys His Gly Lys Cys Pro Phe Pro Leu Ala Asp His

                        ███████
                    Leu Asn Gly Ser Asn His Ala Ile Leu Gln Thr Leu Val His Ser Ile Glu Pro Glu Asp
                    Leu     Asp Thr Gln Tyr Ser Lys Val Leu Ala Leu Tyr Asn Gln His Asn Pro Gly Ala
                    Ser Gly Ser Ser Leu Ser Phe His Ser Thr Val Ile Asn His Tyr Arg Met Arg Gly His
                    Phe Asn Ser Thr Asn His Ala Val Val Gln Thr Leu Val Asn Asn Asn Asn Pro Gly Lys

                                                     *   *
                    Ile Pro                 Leu Pro Cys Cys Val Pro Thr Lys Met Ser Pro Ile Ser Met
                    Ser Ala                 Ala Pro Cys Cys Val Pro Gln Ala Leu Glu Pro Leu Pro Ile
                    Ser Pro Phe Ala Asn Leu Lys Ser Cys Cys Val Pro Thr Lys Leu Arg Pro Met Ser Met
                    Val Pro                 Lys Ala Cys Cys Val Pro Thr Gln Leu Asp Ser Val Ala Met

                    Leu Phe Tyr Asp Asn Asn Asp Asn Val Val Leu Arg His Tyr Glu Asn Met Ala Val Asp
                    Val Tyr Tyr Val     Gly Arg Lys Pro Lys Val Glu Gln Leu Ser Asn Met Ile Val Arg
                    Leu Tyr Tyr Asp Asp Gly Gln Asn Ile Ile Lys Lys Asp Ile Gln Asn Met Ile Val Glu
                    Leu Tyr Leu Asn Asp Gln Ser Thr Val Val Leu Lys Asn Tyr Gln Glu Met Thr Val Val

                         *       *
                    Glu Cys Gly Cys Arg
                    Ser Cys Lys Cys Ser
                    Glu Cys Gly Cys Ser
                    Gly Cys Gly Cys Arg
```

FIGURE 3-11. Comparison of Vg1 protein with selected members of the TGF-β gene family. The probable processing sites are enclosed in boxes, the conserved cysteine residues are marked by asterisks, and the possible glycosylation site conserved between Vg1 and dpp is indicated by the bar. In each case only the carboxyl end of much larger protein is shown. (From *Cell*, 51, 864, 1987. Copyright by Cell Press. With permission.)

Mesoderm Induction in Amphibians: The Role of TGF-β2-Like Factors

F. Rosa, A. B. Roberts, D. Danielpour, L. L. Dart, M. B. Sporn, and I. B. Dawid

Science, 239, 783—785, 1988 3-12

To examine the mesoderm-inducing potential of the mammalian growth factor transforming growth factor (TGF), *Xenopus* animal cap cells were exposed to medium containing either human TGF-beta1 or human TGF-beta2.

Although TGF-beta1 had no effect, TGF-beta2 induced elongation of and expression of alpha-actin mRNA in the animal explants. TGF-beta2 was active as a mesoderm inducer in concentrations from 3 to 200 ng/ml. However, the maximal alpha-actin mRNA level induced by TGF-beta2 was significantly lower than that found in explants treated with media conditioned by the XTC cell line (XTC-CM). There was no synergistic induction effect between TGF-beta2 and FGF. However, there was a synergistic effect between FGF and TGF-beta1. To determine whether the inducing activity in acid-activated XTC-CM was related to TGF-beta, a competitive radioreceptor binding assay was used. XTC-CM competed weakly for the binding of ^{125}I labeled TGF-beta1. Acid-activated XTC-CM stimulated the formation of colonies of NRK cells and inhibited the growth of CCL64 cells, as does TGF-beta. Finally, polyclonal antibodies to porcine TGF-beta2 inhibited the activity of XTC-CM by 80%. Antibodies to TGF-beta1 did not inhibit this activity.

TGF-beta2 is a mesoderm inducer, while TGF-beta1 is not. The potent mesoderm inducer, acid-activated XTC-CM, appears to contain TGF-beta-like activity. This activity is inhibited by anti-TGF-beta2 polyclonal sera. Therefore, a molecule related to TGF-beta2 is probably important in induction of mesoderm in the *Xenopus* embryo.

♦These investigators made the insightful observation that certain properties of Smith's XTC factor resembled properties of TGFβ. They have confirmed their suspicions by demonstrating that antibody specific for TGFβ2 selectively interferes with the α-actin-inducing capability of diluted XTC conditioned medium. They also showed that XTC conditioned medium has activities on mammalian tissue culture cells that resemble those of TGFβ. This paper conflicts with the findings by Kimelman and Kirschner (*Cell,* 51, 869—877, 1987) that pure TGFβ2 has no inducing effect, and FGF must be included for α-actin induction. Rosa et al. find that TGFβ2 alone can induce α-actin, and do not observe the synergistic effect of FGF and TGFβ. *Thomas D. Sargent*

The Differing Effects of Occipital and Trunk Somites on Neural Development in the Chick Embryo

T. M. Lim, E. R. Lunn, R. J. Keynes, and C. D. Stern
Development, 101, 525—533, 1987 3-13

In the peripheral nervous system of vertebrates, axons, neural crest cells, and dorsal root ganglia grow through the cranial half of each sclerotome. However, in the occipital sclerotomes, dorsal root ganglia do not form. The migration of neural crest and the development of sensory ganglia was investigated in the occipital sclerotomes of the chick embryo.

The monoclonal antibody, HNK-1, is a marker for neural crest cells. Immunohistochemical staining indicated that neural crest cells entered the cranial half of the occipital sclerotomes. Zinc iodide/osmium tetroxide staining demonstrated axon growth within these sclerotomes. However, by stage 23 there were no dorsal root ganglia within the occipital sclerotomes. Motor axons did however, grow from these segments. Quail occipital neural tube was grafted in place of chick trunk tube in chick embryos. In this new position, the quail occipital tube developed dorsal root ganglia.

Therefore, failure to develop dorsal root ganglia in the occipital position is not due to differences in the neural tube or crest in the occipital region. It is possible that factors necessary for dorsal root ganglion development, while present in the other sclerotomes, are missing at the occipital level in chick embryos.

♦Both motor axons and neural crest cells migrate through the rostral half of each somite in the trunk region of avian embryos. This metameric pattern of migration leads to the segmentation of the dorsal root ganglia and ventral roots and is, therefore, a vital component for proper development of the peripheral nervous system. The region containing the first five somites (called occipital somites) differs from other axial levels in that no dorsal root ganglia form therein. Does this represent an inherent property of these somites or is it intrinsic to the neural crest cells derived from this region?

Here, Lim et al. examine the migration of neural crest cells and motor axons in occipital regions. They find that both neural crest cells and axon move through the rostral half of each somite, as they do in the trunk. Furthermore, when neural tube plus neural crest from the occipital region are transplanted into the trunk, they give rise to apparently normal dorsal root ganglia. These results indicate that the neural crest cells migrate similarly through the trunk and occipital somites and have similar differentiative capacity to form dorsal root ganglia. The logical conclusion is that the environment in the occipital regions lacks a growth or survival factor which permits development of the dorsal root ganglion cells. Thus, different axial regions can utilize the same migratory information and share common developmental potentials, yet give rise to different derivatives. The different patterns of cellular differentiation may be generated by region-specific distribution of certain molecules (such as survival factors?) on an otherwise shared environment. *Marianne Bronner-Fraser*

The Same Inducible Nuclear Proteins Regulates Mitogen Activation of Both Interleukin-2 Receptor-Alpha Gene and Type 1 HIV

E. Böhnlein, J. W. Lowenthal, M. Siekevitz, D. W. Ballard, B. R. Franza, and W. C. Greene

Cell, 53, 827—836, 1988 3-14

The growth of human T cells depends on mitogen- or antigen-induced expression of interleukin-2 (IL-2) and IL-2 receptors (IL-2Rs). Activation of the long terminal repeat (LTR) of type 1 human immunodeficiency virus (HIV-1) resembles activation of the IL-2R alpha promoter. The *cis*-acting sequence and the *trans*-acting factor that regulate mitogen inducibility of the IL-2R alpha promoter were investigated, and their effects on the HIV-1 LTR were examined.

To identify factors that bind to the IL-2R alpha promoter, nuclear extracts from stimulated and unstimulated Jurkat T cells were incubated with ^{32}P end-labeled IL-2R oligonucleotide probes (see Figure 1, *Cell*, 53, 828, 1988). Factors from the induced T cells bound to the IL-2R alpha promoter region. Synthetic mutants of this region were constructed and used in competition and direct binding experiments to define more narrowly the region important for binding. The binding site was a 12-base pair palindromic sequence between nucleotides -267 and -256 (see Figure 3-14). This sequence was homologous to the transcriptional enhancer of the HIV-1 LTR. Competition experiments indicated that the HIV sequence could block all binding to the IL-2R alpha promoter. Microscale DNA-affinity precipitation assays were performed, and specific binding of an inducible 86-kDa protein, HIVEN86A, was detected with both the IL-2R alpha promoter and HIV-1 LTR regions. Both of these binding sites could be used to confer mitogen inducibility in either orientation on a heterologous TK promoter linked to the CAT gene.

These results demonstrate that the inducible protein HIVEN86A specifically associates with the IL-2R alpha promoter and the HIV-1 enhancer region. These binding sites are sufficient to confer mitogen inducibility on a heterologous promoter. This indicates that HIV-1 could subvert the regulatory system of the IL-2R alpha gene to activate its own LTR and enhance viral transcription.

♦Antigen or mitogen activation of immunocompetent T lymphocytes results in the *de novo* expression of cellular genes, notably the interleukin 2 (IL2) and its receptor IL2R (an $\alpha\beta$ heterodimeric structure with high affinity for its ligand) genes which encode products required for T cell growth and

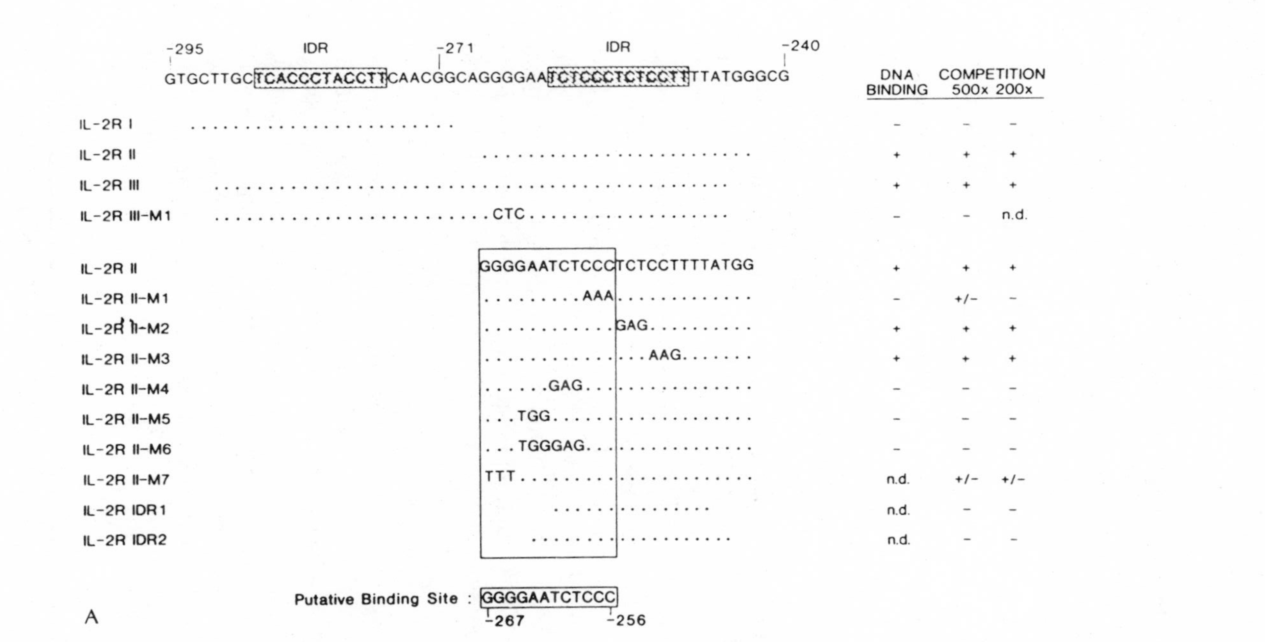

FIGURE 3-14. Summary of IL-2Rα promoter binding data. (A) Summary of competition and direct binding results using mutant IL-2Rα oligonnucleotides. The sequence of the IL-2Rα promoter region between nucleotides -295 and -240 is shown; the two IDRs are indicated by cross-hatched boxes. Sequence identity of the oligonucleotides to the wild-type sequence is represented with dots; mutations are shown by substituted nucleotides. Results for each oligonucleotide in the direct binding and competition assays are at right. The putative 12-bp protein binding site deduced from these studies is shown. (B) Sequence homology (9/11 bases) between the IL-2Rα promoter protein binding site and 3′ repeat of the HIV-1 enhancer. The general organization of the HIV-1 LTR is also depicted, including the location of the direct repeats (DRs) transcriptional enhancer, Sp1 binding sites, TATA box, transcription initiation site (+1), and the *trans*-acting responsive region (TAR) involved in *tat* HIV-1 action. Sequences of synthetic oligonucleotides used in subsequent experiments corresponding to either both HIV-1 DRs (HIV-DR) or the 5′ (HIV-R5′) or 3′ (HIV-R3′) element are shown. Lowercase letters indicate linker-derived nucleotides. (From *Cell*, 53, 831, 1988. Copyright by Cell Press. With permission.)

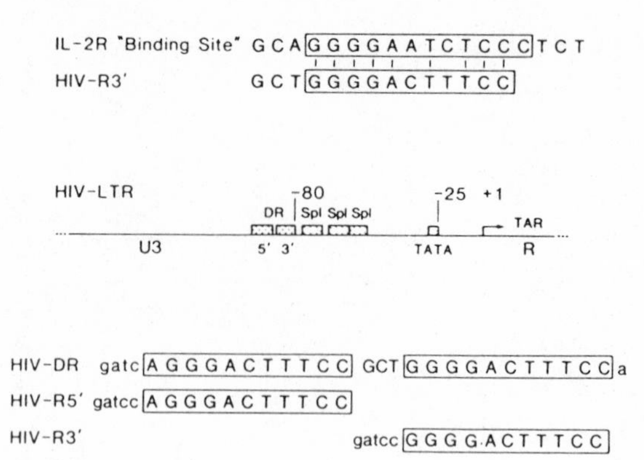

FIGURE 3-14B

proliferation The regulatory constraints on this autocrine growth pathway, particularly the nuclear factors involved, are poorly understood but are important not only in the development of "normal" immune responses but in the progression of AIDS. The activation of T cells latently infected with HIV-1 also enhances viral gene expression and consequent viral replication and must be an important contributor to the pathophysiology of the disease. Resolution of the molecular basis of this enhancement may be key to improving the survival prospects of AIDS victims. Böhnlein et al. have identified a *cis*-acting enhancer element in the 5′ regulatory region of the IL2Rα (p55) gene which is highly homologous to the enhancer elements present in the HIV-1 LTR (long terminal repeat), located within the U3 region of the viral genome, and have shown that the same 86 kDa DNA-binding nuclear protein of T cells binds to both the IL2Rα and HIV-1 enhancer elements. The enhancer elements were mapped, using mutated oligonucleotides and gel retardation assays in which binding was assessed directly and by competition. The 86-kDa nuclear protein was partially purified by the DNA-affinity precipitation technique. The functionality of the enhancer elements was indicated by the fact that, in either orientation, they conferred mitogen inducibility on a plasmid consisting of the TK promoter linked to the CAT gene and transfected into mitogen-responsive Jurkat T lymphocytes. These data suggest that HIV-1 can subvert the normal action of an activation-induced nuclear protein to augment viral gene transcription. *Judith A. K. Harmony*

Mouse Embryonic Stem Cells Exhibit Indefinite Proliferative Potential

Y. Suda, M. Suzuki, Y. Ikawa, and S. Aizawa
J. Cell. Physiol., 133, 197—201, 1987 3-15

Segregation of germ from somatic cells occurs at approximately 7 d of gestation in the mouse. To determine whether early embryonic stem (ES) cells prior to this separation are immortal, the proliferative potential of day 4 ES cells was examined.

Only one out of eight ES lines examined was diploid. These cells were propagated and after ten passages, pseudodiploid cells became evident. However, diploid lines could be cloned from these cells. No signs of crisis or spontaneous transformation were observed after an average of 250 cumulative doublings. These cells retained a normal karyotype. The ES cell line was successfully used to generate chimeric mice. These mice were normal, fertile, and did not develop tumors.

These results suggest that these very early ES cells are immortal without transformation.

♦These investigators have examined carefully the proliferative potential of *in vitro* cultured embryo-derived stem (ES) cells. If these cells were prone to crisis and spontaneous transformation, they would not be useful for the production of mutant mice. In fact, it was reported in this paper that only one out of eight newly established lines possessed a diploid karyotype. Although this is low, the one line that was diploid could be maintained in culture for up to 250 cumulative doublings with no crisis or transformation occurring. A recloning of diploid cells, however, was necessary after every 100 doublings. The cells were considered to be normal at the termination of the experiment because they possessed a diploid karyotype and were able to form germline chimeras without causing tumors. These results indicate that ES cells are immortal, and if a stable diploid line is identified at the start of an experiment, it should be possible to retain the karyotype during *in vitro* manipulations. This paper is also important because it emphasizes that not all ES lines possess a diploid karyotype, including newly established lines, and that it is necessary to monitor the karyotype of any line before considering its use for blastocyst-injection chimeras. *Terry Magnuson*

Maternal Controls, Cytoplasmic Determinants and Imprinting

4

INTRODUCTION

Following fertilization of an egg, cleavage divisions may be either symmetric or asymmetric in the developing embryo. Since the vast majority of cytoplasm in the embryo is of maternal origin, maternal factors directly influence embryonic events.

Genetic "factors", as compared to biochemical maternal factors, can also impact on embryonic events, not so much by what a gene encodes, but whether it is "preprogrammed" to express at all. Thus, "genetic imprinting" in one generation will effect expression in the next.

The articles in this chapter deal directly with the biochemical nature of cytoplasmic determinants and how cellular events can influence their subsequent distribution. In addition, the articles discuss specific maternal genes that impart localization signals in the embryo, and possible mechanisms of genetic imprinting that can influence gene expression one generation later.

Identification of Genes Required for Cytoplasmic Localization in Early C. elegans Embryos
K. J. Kemphues, J. R. Priess, D. G. Morton, and N. Cheng
Cell, 52, 311—320, 1988 4-1

The early embryonic blastomeres of the nematode *Caenorhabditis elegans* exhibit marked differences in behavior during development. Qualitative differences—not the absolute amount of cytoplasm—are apparently responsible for these distinctive behaviors. The well-developed genetics of *C. elegans* allows a mutational analysis of cytoplasmic localization in early development. The authors have used a new screening method to isolate nonconditional maternal effect lethal mutants, and have examined these for defects in early development.

The eight strict maternal effect mutations isolated identify four genes (*par-1, par-2, par-3, par-4*) that are required for cytoplasmic localization in early embryos of *C. elegans.* Mutations in the genes produce defects in patterns of cleavage, timing of cleavages, and localization of germ line-specific P granules. Four mutations in *par-1* and *par-4* are fully expressed maternal effect lethal mutations. Embryos from homozygous mothers arrest as masses of differentiated cells lacking intestinal cells. Four mutations in *par-2*, *par-3*, and *par-4* are incompletely expressed maternal effect lethal mutations. Some embryos from homozygous mothers grow to infertile adults because of the absence of functional germ cells.

All these defects presumably reflect the failure of a maternally encoded system for intracellular localization in early embryos. The germ line appears to be particularly sensitive to mutations in the *par* genes. While the *par* genes do not encode actin proteins, they may encode proteins required for the functions or distribution of actin microfilaments.

♦The first division of the *C. elegans* zygote is asymmetric and the fates of the resulting two daughter cells are very different. These fates are believed to be controlled by the asymmetric distribution of one or more cytoplasmic determinants during this division. An example of such a putative determinant are the P granules which get segregated reproducibly to one daughter cell at each of the first four embryonic divisions to end up in the germ line precursor P4. This paper described the isolation and characterization of mutations in four genes required for asymmetric position and orientation of the mitotic spindle in the first division(s) and for the proper segregation of the P granules (and possibly other cytoplasmic determinants as well). Strome and Wood (*Cell*, 35, 15—25, 1983) showed previously that the asymmetric position of the first division and the proper segregation of P granules could be blocked with cytochalasin, implicating actin microfilaments in these processes. The *par* genes are not linked to the known actin genes indicating that they do not encode actin proteins. They likely encode products required for specialized functions of microfilaments or for their distribution in early embryos. Molecular analysis of the *par* gene products should provide major insights into the mechanisms used in establishing asymmetries during the first divisions of the embryo. *Joseph G. Culotti*

An Analysis of the Role of Microfilaments in the Establishment and Maintenance of Asymmetry in *Caenorhabditis elegans* Zygotes

D. P. Hill and S. Strome

Dev. Biol., 125, 75—84, 1988 4-2

In many systems, specification of cell destiny is controlled by an unequal partitioning of cytoplasmic components in the egg. The authors used the asymmetrical first division of *Caenorhabditis elegans* embryos in an attempt to elucidate the process regulating the distribution of developmental potential. In this system, microfilaments are required to generate asymmetry. The microfilament inhibitor cytochalasin D served to disrupt microfilament structure at different times in the first cell cycle.

Microfilaments were found to be necessary only for a brief time, about three fourths of the way through the first cell cycle, for manifestations of asymmetry to occur. The events affected are pseudocleavage, pronuclear migration, germ-granule segregation, and movement of the mitotic spindle to an asymmetric position. Microfilaments were not required to maintain asymmetries that were already established.

A critical interval was defined during which proper microfilament structure is needed for events occurring during the interval and afterward. Asymmetric placement of the mitotic spindle actually takes place subsequent to the critical interval. Presumably actin filaments provide spatial cues or participate in events that ultimately determine the position of the mitotic spindle.

♦This paper provides some of the first insights into the molecular mechanisms used to segregate cytoplasmic determinants to one of the two daughter cells generated by the first embryonic division.

Hill and Strome show that acting microfilaments (which are required for the segregation of P granules to one daughter cell during the first four division of the *C. elegans* embryo) act at the time when the first asymmetries become apparent during the first embryonic division, not before and not after. Microfilaments are not required to maintain the asymmetric distribution of P granules once they have been asymmetrically segregated. *Joseph G. Culotti*

Identification of a Microtubule-Based Cytoplasmic Motor in the Nematode C. elegans

R. J. Lye, M. E. Porter, J. M. Scholey, and J. R. McIntosh

Cell, 51, 309—318, 1987 4-3

Because *Caenorhabditis elegans* lacks axonemal dyneins, can grow in liquid culture at high density, and is suitable for molecular genetic and morphologic analysis, it is well suited for studies of the role of cytoplasmic dyneins in cellular transport. Microtubules and associated proteins were prepared from extracts of *C. elegans*. From these was isolated a high-molecular-weight protein that has some dynein-like properties and also has *in vitro* motility characteristics of kinesin. The microtubule stabilizer taxol was used to stimulate polymerization of microtubules in extracts of gravid adult nematodes.

The microtubule binding protein had MgATPase activity and copurified with a polypeptide of M_r, about 400 kDa. The ATPase activity was 50% inhibited by 10 μM vanadate or 1 mM N-ethyl maleimide. It was enhanced by Triton. The polypeptide was cleaved at a single site by UV light in the presence of ATP and vanadate. The protein promoted ATP-dependent translocation of microtubules or axonemes. Motility was blocked by 5 μM vanadate and by *N*-ethyl maleimide.

This protein would seem to be a novel microtubule translocator having features of both dynein and kinesin. It is not likely associated with an axoneme. Polyclonal antibodies will help demonstrate the function of this polypeptide *in vivo.*

♦This paper represents only the second attempt to biochemically characterize microtubule associated proteins in *C. elegans* (see also Aamodt and Culotti, *J. Cell Biol.*, 103, 23—31, 1986).

This is an important paper because it not only identifies what is probably a novel microtubule associated "motor" with properties of both "kinesin" and "dynein", but it also does it in an organism amenable to genetic analysis. *C. elegans* may be the most promising genetically amenable invertebrate for studying the functions of different microtubule-associated proteins, because its simple anatomy should allow one to localize these proteins to specific, identifiable cells. Genetics can then be applied to understand the role that these proteins play in microtubule function *in vivo,* and possibly to identify interacting proteins by revertant analysis. *Joseph G. Culotti*

Determination of Cell Division Axes in the Early Embryogenesis of *Caenorhabditis elegans*

A. A. Hyman and J. G. White

J. Cell Biol., 105, 2123—2135, 1987 4-4

Specific cytoplasmic components of *Caenorhabditis elegans* may help determine the differentiated state of early blastomeres. The authors analyzed movement of the centrosomes in the early embryonic divisions of *C. elegans* in order to determine how the division axes are established. In one set of blastomeres, the division axes are established successively on the same axis and the blastomeres divide unequally. Division axes in the other set are established in an orthogonal pattern, and the sister blastomeres have similar developmental potential.

Centrosome separation follows a consistent pattern in all cells, and this in itself leads to an orthogonal pattern of cleavage. In cells dividing on the same axis there is an added directed rotation of pairs of centrosomes along with the nucleus through defined angles. Intact microtubules are necessary for this rotation. Inhibitors of polymerization prevent rotation, as does depolymerization of microtubules. Study of the distribution of microtubules in fixed embryos during rotation suggested that microtubules running from the centrosome to the cortex are important in aligning the centrosome-nuclear complex.

These findings indicate that the rotation of pairs of centrosomes in some cells of *C. elegans* aligns the mitotic apparatus along a segregation axis, facilitating the unequal partitioning of components into daughter cells.

♦This paper examines the details of centrosome and nuclear migration and rotation that account for the determinitive asymmetric divisions of the germline precursor cells during early embryogenesis of *C. elegans*. These details are essential to understand how events such as mitotic spindle orientation and segregation of cytoplasmic determinants may be coordinated. For example, this paper shows that rotation of the centrosome-nuclear complex brings the centrosomes into a predetermined axis, which is the same axis along which the P granules (putative cytoplasmic determinants) migrate. That microtubules have a role in migration and rotation is clear. It is also clear that actin microfilaments are probably required for rotation and for P granule segregation. How these two cytoskeletal systems interact to bring about the precisely coordinated movements of the centrosome-nucleus complex, and P granule migration should be the subject of some interesting research in the future. *Joseph G. Culotti*

Maternal Genes Required for the Anterior Localization of *bicoid* Activity in the Embryo of *Drosophila*
H. G. Frohnhöfer and C. Nüsslein-Volhard
Genes Dev., 1, 880—890, 1987 4-5

The segmented pattern of the *Drosophila* embryo is organized by two terminal centers at the egg poles, both defined by transplantable cytoplasmic activities and by maternal-effect genes. The phenotypes of genes influencing the anterior pattern are quite diverse, and the *bicoid* (*bcd*) locus is phenotypically unique. The *torso* group of genes involve parts of the anterior head region. Embryos mutant for *exuperantia* (*exu*) and *swallow* (*swa*) have pattern defects in the head region that distinguish them from both the *torso* group genes and *bcd*.

Evidence now is available that, in *exu* and *swa* mutant embryos, bcd^+ activity is more evenly distributed than in wild-type embryos, suggesting that both these genes are involved in the anterior localization of bcd^+ activity. The mutant phenotypes of *swa* and *exu* are characterized by a lack of anterior head structures, like weak *bcd* alleles. Cytoplasmic transplantation studies showed much reduced bcd^+ activity at the anterior tip of *swa* and *exu* embryos. In contrast to weak *bcd* alleles, removal of anterior cytoplasm did not much affect the *swa* or *exu* phenotypes.

It appears that, in embryos mutant for *exu* or *swa*, *bcd* activity is not restricted to the anterior tip, but extends posteriorly. These genes might function in the establishment of both polar centers or influence their interactions. Alternately, they could act on *bcd* alone, with this in turn mediating abdominal defects. The present data support the latter concept.

The Role of Localization of *bicoid* RNA in Organizing the Anterior Pattern of the *Drosophila* Embryo

T. Berleth, M. Burri, G. Thoma, D. Bopp, S. Richstein, G. Fregerio, M. Noll, and C. Nüsslein-Volhard

EMBO J., 7, 1749—1756, 1988 4-6

The gene *bicoid* (*bcd*) is responsible for development of the entire anterior half of the *Drosophila* embryo. In embryos from flies homozygous for strong *bcd* alleles, all head and thoracic structures are missing, and the blastodermal fate map of the remaining abdominal segments is shifted anteriorly. The authors cloned and sequenced the *bcd* locus and analyzed its transcripts. Analyses of *bcd* mRNA also were carried out in *exu* and *swa* embryos, in which the most anterior structures are lacking. These phenotypes apparently are caused by a shallow distribution of bcd^+ activity throughout the egg.

The maternal-effect gene *bcd* was identified in an 8.7-kb genomic fragment. The major transcript of 2.6-kb includes a homeobox with low homology to known homeoboxes; a PRD-repeat; and an M-repeat. *In situ* hybridization showed that *bcd* is transcribed in the nurse cells. The mRNA is localized at the anterior tip of the oocyte and early embryo up to the cellular blastoderm stage. Localization of the transcript requires *exu* and *swa*. Transcript stability is lowered by functions dependent on posterior group genes.

In vitro mutagenesis will help demonstrate the structural features of *bcd* mRNA required for localization. The simplest mechanism of localization is formation of a specific RNA-protein complex that is immobilized in the oocyte and embryo by binding to the cytoskeleton of the egg cell. The gene products of *exu* and *swa* may function to form complexes with the *bcd* RNA, trapping it at the anterior end of the oocyte.

♦Development of the *Drosophila* embryo depends on maternal factors residing at the anterior and posterior poles. Three maternal genes code for products that deeply influence the development of anterior structures, *bicoid (bic), exuperantia (exu),* and *swallow (swa).* Mutants in the three genes have similar phenotypes: missing head and thorax are replaced by duplicated posterior structures. Using a combination of genetic, cytoplasmic transplantation/rescue and cytoplasmic removal experiments, Frohnhöfer et al. conclude that one of the roles of *exu* and *swa* gene products is to localize *bic* activity. Strong circumstantial evidence supports this conclusion. For instance, cytoplasmic transplantation/rescue experiments showed that *bic* activity is much reduced in the anterior pole of *exu* and *swa* embryos, even though these embryos were wild type for the *bic* gene. Moreover, *exu/bic* double mutants have only the *bic* phenotype, consistent with the idea that *exu* plays a role in *bic* localization. In the second paper, Berleth et al. demonstrate directly that this hypothesis is essentially correct. They report the cloning of the *bic* gene and the use of molecular probes to localize the mRNA in tissue sections. In wild-type embryos the *bic* mRNA is strongly localized to the anterior tip of oocytes of all developmental stages and of early embryos. Significantly, and as predicted by the previous studies, in *exu* and *swa* mutants *bic* mRNA has normal abundance but is found distributed along the entire length of the oocytes and embryos.

Although an asymmetric localization of gene products in the *Drosophila* embryo has been demonstrated in a few other cases, these two reports are significant because they identify genes that may be involved in the actual localization and because they provide a basis for generating hypotheses on the possible mechanisms involved in the localization. Specifically, *exu* and *swa* may be part of a bridge that connects *bic* mRNAs to the oocyte cytoskeleton. *Marcelo Jacobs-Lorena*

Involvement of the *pumilio* Gene in the Transport of an Abdominal Signal in the *Drosophila* Embryo

R. Lehmann and C. Nüsslein-Volhard

Nature (London), 329, 167—170, 1987 4-7

Five genes are required for abdominal segmentation and pole cell formation posteriorly in the *Drosophila* embryo. Cytoplasmic transplant studies with *oskar* (*osk*) mutant embryos suggest that the pole plasm is the source of a signal required in the anterior abdominal region. Evidence has been obtained that the maternal gene *pumilio* (*pum*) is involved in transport of this signal to the abdominal region.

Embryos derived from females homozygous for a strong *pum* allele form no more than two of the normal eight abdominal segments. The authors determined whether the abdominal phenotype of *pum* embryos can be rescued by transplanting wild-type cytoplasm. The pole plasm is normal in *pum* embryos. Transplant studies showed that *pum* pole plasm contains the abdominal signal, but that it does not reach the abdominal target site. Abdominal segmentation occurred when physical separation of the pole plasm and abdominal region was overcome, either by transplanting *pum* pole plasm into the abdominal region of *pum* embryos, or genetically in double mutants where the abdominal region is juxtaposed to the pole plasm.

The *pum* gene product does not function only in early embryogenesis. The gene may be transcribed both maternally and during larval or pupal development, in the latter instance for imaginal disk development. The *pum* gene may be one of those genes that is not required exclusively for a given developmental process, but is necessary for it to take place.

♦The maternally expressed gene *pumilio (pum)* is required for the proper development of abdominal structures of the *Drosophila* embryo. Making use of very elegant cytoplasmic transplantation experiments, the authors show that the *pum* gene product is involved in the transport of an "abdominal signal" from the extreme posterior pole of the embryo to a more anteriorly located position, where the prospective abdomen develops.

Only a few maternal genes, among them *pum* and *oskar (osk),* are required for the normal patterning of the posterior end of the *Drosophila* embryo. Embryos derived from females homozygous for a strong *pum* allele form no more than two of the normal eight abdominal segments while more anterior structures of the embryo develop normally. The *pum* phenotype can be rescued by injection of wild-type cytoplasm. Interestingly, the rescuing activity is tightly localized to the very posterior end; rescue is obtained only if posterior polar cytoplasm [0% egg length (EL)] is injected into the prospective abdominal region (20 to 40% EL). Injection of abdominal wild-type cytoplasm into the mutant's abdominal region is completely ineffective. These and other experiments suggested that the *pum* phenotype is not due to the absence of a factor required for proper development of embryonic abdominal structures, but simply due to the failure of transporting this factor from the end of the embryo (0% EL) to the prospective abdominal area (20 to 40% EL). The authors argued that if this were true, it should be possible to rescue the *pum* phenotype simply by injecting *pum* polar cytoplasm into the abdominal region of *pum* recipients. This is the equivalent of mechanically "transporting" the factor for the embryo. The outcome of this experiment was clear; transplantation of 0% EL *pum* mutant cytoplasm into 20 to 40% EL *pum* mutant embryos rescued the phenotype, thus confirming the hypothesis. Another set of experiments made use of mutants that change the fate map of the embryo. In *torso-like (tsl)* mutants, the abdomen is derived from the very posterior end of the embryo rather than the usual 20 to 40% EL region. The authors argued that in this situation the need to transport the factor from the pole to the prospective abdominal region should be obviated. This is exactly what they found. In *pum/tsl* double mutants the abdomen develops properly at the posterior pole, without any cytoplasmic transplantations. From other evidence presented in the paper, the authors speculate that the *pum* gene product may be associated with the microtubule network to perform the transport function and that the signal transported by *pum* corresponds to the *osk* gene product.

This is an important paper that bears on the issue of cytoplasmic localization and transport of embryonic determinants. The experiments demonstrate the power of combining the rich genetic background of *Drosophila* with classical cytoplasmic transplantation experiments. *Marcelo Jacobs-Lorena*

An RNA-Binding Protein from *Xenopus* Oocytes is Associated with Specific Message Sequences

D. R. Crawford and J. D. Richter

Development, 101, 741—749, 1987 4-8

Early development in *Xenopus* is relatively dependent on maternal mRNA. RNA-binding proteins may be responsible for the regulated expression of some mRNAs in early development. The authors have obtained evidence that a particular RNA-binding protein interacts with specific mRNAs in *Xenopus* and that the mRNAs enter polysomes at stages of development where the binding protein is present in lesser amounts.

Monoclonal antibodies against RNA-binding protein from *Xenopus* oocytes were used to immunoselect messenger ribonucleoprotein (mRNP) particles. RNA from this source served to direct the synthesis of oligo (dT)-primed ^{32}P-cDNA. Preparations of cDNA from immunoselected and unselected fractures were used to probe *Xenopus* stage-1 oocyte cDNA libraries. Three cDNA clones were derived specifically from antibody-selected mRNPs. In very early oogenesis, RNA-binding protein and the coselected mRNAs sedimented in the nontranslating mRNP region of a sucrose gradient. Later, the binding protein concentration fell substantially relative to polyadenylated RNA. At this stage the mRNAs were present mainly in the polysome region of a sucrose gradient.

Xenopus oocytes contain an RNA-binding protein, p56, that binds specific message sequences, and may regulate their expression. There probably are broad classes of proteins which bind several different mRNAs. This could be demonstrated using antibodies against different developmentally regulated mRNA proteins.

♦Translational control mechanisms are likely to be critical in early embryonic development, especially in organisms such as *Xenopus* that have large maternal mRNA pools, and a period of rapid cleavage during which there is minimal transcription. Unfortunately, not much is known about proteins that interact in a sequence-specific manner with mRNAs. This paper reports some important progress in this direction; they have used a monoclonal antibody that precipitates a protein called p56 which is a member of a class of oocyte-specific RNA-binding proteins. Precipitated RNP particles from oocyte were used to make a cDNA probe which was used to screen a cDNA library. Several different clones were obtained representing mRNAs that are bound to p56, and translationally repressed, in the early oocyte. When p56 levels decline by oocyte stage VI, these RNAs are being translated efficiently, and it may well be that p56 has a functional role in the repression of these mRNAs. *Thomas D. Sargent*

Axis and Germ Line Deficiencies Caused by U.V. Irradiation of *Xenopus* Oocytes Cultured *in vitro*

S. Holwill, J. Heasman, C. R. Crawley, and C. C. Wylie
Development, 100, 735—743, 1987 4-9

Early development is to some degree controlled by the spatial arrangement of molecules in the oocyte. UV irradiation of the vegetal pole of the fertilized egg of *Xenopus laevis* affects both the development of the embryonic dorsal axis and the formation of primordial germ cells. A dose-dependent reduction of the dorsal axis follows irradiation of the vegetal hemisphere of the egg in the first cell cycle. Similar effects occur with cold and pressure.

The authors developed a method of routinely maturing and fertilizing cultured oocytes of *Xenopus*, in order to study the role of cytoplasmic localization in the oocyte. It was found that the UV-sensitive time for primordial germ-cell and dorsal-axis formation extends back into stage-6 oocytes, indicating a true contribution of the oocyte to these processes.

This is the first instance where experimental manipulation of oocytes has produced a specific developmental effect. The *Xenopus* oocyte is as sensitive as the egg to UV damage of the germ line. UV-induced damage is inheritable from the stage-6 oocyte. Future studies will identify UV-sensitive components in the oocyte. This approach permits study of the roles of many oocyte components in early development.

♦Because of the rapid cleavage cycles during the blastula stage of *Xenopus*, there is very little mRNA synthesis until just before gastrulation. Nevertheless, a lot happens during these first few hours, including mesoderm specification. Maternal control mechanisms are likely to be important in these processes, and the technique for maturing and fertilizing oocytes described in this paper opens up many possibilities to disrupt, augment, and otherwise modify the oocyte and then test the effects on development. The authors also show, using their method, that UV irradiation of oocyte vegetal poles has an effect similar to that of irradiating fertilized eggs, i.e., suppression of dorsal axis formation, and sterility. *Thomas D. Sargent*

Genomic Imprinting Determines Methylation of Parental Alleles in Transgenic Mice

W. Reik, A. Collick, M. L. Norris, S. C. Barton, and M. A. Surani
Nature (London), 328, 248—251, 1987 4-10

In mouse embryogenesis, both the maternal and paternal genomes are required for development to term.

Specific modifications may be imprinted onto the chromosomes during gametogenesis. They are stably propagated, and their expression results in complementary contributions by the parental genomes to development. A substantial part of the genome could be subject to chromosomal imprinting. The molecular nature of the imprinting process is not understood.

The authors used random DNA insertions in transgenic mice to probe the genome for modifications. DNA methylation patterns of transgenic alleles were compared after transmission from the mother or father in seven murine strains carrying autosomal insertions of the same transgenic marker, a chloramphenicol acetyl transferase (CAT) gene linked to an Ig heavy-chain enhancer. One locus showed a clear difference in DNA methylation specific for its parental origin. The parternally inherited copy was relatively undermethylated. The methylation pattern was reversed on each germ-line transmission to the opposite sex. In the other six strains, the transgene was highly methylated in the embryo.

These findings suggest heritable molecular differences between maternally and paternally derived alleles on mouse chromosomes. Some chromosomal regions must maintain imprints that distinguish parental alleles until late in development. There is increasing evidence that genomic imprinting is important in the control of mammalian embryogenesis. It may also influence the phenotypic expression of genetic disorders in the human.

Degree of Methylation of Transgenes is Dependent on Gamete of Origin

C. Sapienza, A. C. Peterson, J. Rossant, and R. Balling
Nature (London), 328, 251—254, 1987 4-11

Differences in function of the maternal and paternal genetic contributions to the mammalian zygote nucleus are ascribed to genome imprinting during male and female gametogenesis. Any imprint seemingly should be physically linked with the pronucleus and persist through DNA replication and cell division. The mechanism of imprinting must be able to influence gene expression and be capable of changing the gender identity of the imprint.

Differential DNA methylation during gametogenesis or before formation of the zygote nucleus could satisfy the first three of these criteria. It now appears that the methylation patterns of exogenous DNA sequences in transgenic mice can be altered by switching their gamete of origin in successive generations. Methylation status of transgene loci was studied in five lines hemizygous for the quail fast fiber muscle-specific troponin I (TNI) gene. In four of the five lines, the methylation state of the transgene varied with the gamete of origin.

A DNA methylation imprint imposed during gametogenesis can persist into the adult and alter gender identity in successive generations. If such a methylation-based imprinting mechanism operates on endogenous loci, it could explain quantitative variations in phenotype that are not reflected in genotype.

Maternal Inhibition of Hepatitis B Surface Antigen Gene Expression in Transgenic Mice Correlates with *de novo* Methylation

M. Hadchouel, H. Farza, D. Simon, P. Tiollais, and C. Pourcel
Nature (London), 329, 454—456, 1987 4-12

Methylation of DNA may explain differential modifications of the genome taking place during gametogenesis, which result in functional differences between the paternal and maternal genomes at the time of fertilization. Two transgenic male mice have been described which contain hepatitis B viral sequences in their genomes and express hepatitis B surface antigen.

The authors describe a transgenic mouse strain in which expression of HBsAg is irreversibly repressed after passage through the female germ line. Inhibition is associated with methylation of all HpaII and HhaI sites within the foreign gene, which are integrated into a site on chromosome 13.

The irreversibility of HBsAg gene expression in this system contrasts with other transgenic mouse sequences which are reversibly methylated after passage through the male or female germ line. In either instance, however, methylation is important in the imprinting process. Demethylation may occur in oocytes, followed by *de novo* methylation of the female allele only in the early embryo.

Parental Legacy Determines Methylation and Expression of an Autosomal Transgene: A Molecular Mechanism for Parental Imprinting

J. L. Swain, T. A. Stewart, and P. Leder

Cell, 50, 719—727, 1987 4-13

There is mounting evidence from studies in the mouse that the expression of certain genes is determined by inheritance from the male or female parent. The concept of parental imprinting is supported by data showing that both genomes are required for normal embryonic development. In the course of studies of the *in vivo* effect of an activated *c-myc* oncogene, a transgenic mouse was produced in which the transgene is autosomally inherited; its expression is determined solely by its parental origin.

The autosomal transgene bears elements of the Rous sarcoma virus (RSV) LTR and a translocated *c-myc* gene. If the transgene is inherited from the male parent, it is expressed only in the heart. If from the female parent, it is not expressed at all. This pattern correlates with a parentally imprinted methylation statue that is evident in all tissues. Methylation of the *trans*-gene is acquired by passage through the female parent and eliminated during gametogenesis in the male.

Apparently, autosomal gene expression can depend on the parent from which it is inherited. The parental imprinting effect may be easily damaged or "leaky" property of the *trans*-gene. The findings may help explain the apparent non-Mendelian behavior of some autosomal genes. Juvenile Huntington's disease and infantile myotonic muscular dystrophy may represent human manifestation of parental-specific gene expression.

♦These four papers make use of transgenes that have randomly inserted into the mouse genome to study possible molecular mechanisms responsible for imprinting. In these studies, the methylation of CpG dinucleotides and the expression of maternally or paternally derived transgenes was examined. The overall findings can be summarized as follows: in all cases except one, methylation differences were observed such that the transgene was hypermethylated when transmitted through the female germ line but undermethylated when passed through the male germ line. In the one exception, the results were reversed (Sapienza et al., 1987). Although expression was detected with a couple of different transgenes (Sapienza et al., 1987; Swain et al., 1987), only one showed an expression pattern that correlated with hyper- or hypomethylation (Swain et al., 1987). The degree of methylation, however, was uniform in all tissues regardless of whether tissue-specific expression was observed. In three of the reports, the imprinting or state of methylation of the transgene could be reversed upon transmission through the germline of the opposite sex (Reik et al., 1987; Sapienza et al., 1987; Swain et al., 1987). In one case, however, the transgene was irreversibly hypermethylated when passed through the female germline (Hadchouel et al., 1987). Only one of the insertions sites has been mapped, and it was found to be located on a chromosome which according to the earlier genetic studies shows no deleterious effects of imprinting. Questions that remain from these studies center on the reason for the high frequency of transgenes showing germline modification. Is this specific to transgene sequences that are susceptible to imprinting, or is it due to preferred sites of integration into imprinted chromosomal regions? The fact that little correlation exists between methylation and expression detected in the above studies also raises the question of what exactly constitutes the primary imprinting signal. Does methylation play a functional part in this process, and if not, what are the molecular mechanisms involved. Other points of interest involve determining when the imprinting process occurs and how long during development does the primary signal exist. The phenomenon of imprinting in mammals is one of the more fundamental concepts to emerge from recent work and the above papers have begun to address some of the key issues regarding the molecular mechanisms responsible for this process. *Terry Magnuson*

Cell Interactions 5

INTRODUCTION

The manner in which individual cells behave within a cell population is, to a large degree, influenced by the associations made with other cells and with extracellular material. Thus, the nature of cell interactions has important ramifications in the progress of developmental programs.

One obvious level of cell interaction is between neighboring cells within a cell population. During embryogenesis in most organisms, cell movement plays a crucial role. Cells must be able to adhere to each other, disengage, and, following migration, readhere. The appropriate expression of cell surface molecules that mediate cell-cell adhesion is critical. The molecular nature of such molecules has been the focus of extensive investigation.

In addition to other cells, individual cells within a multicellular organism interact with the extracellular matrix. This interaction is particularly important during cell movement in directing the appropriate pathway that cells should follow.

Once cells have migrated to the appropriate embryonic location and adhered to their nearest neighbor, cell-cell communication becomes an important function. Such communication is critical for assuring the correct future behavior of a particular cell, whether it will be induced to differentiate into one particular cell phenotype or another.

The articles in this chapter offer a broad look at several aspects of developmental cell interactions. The articles discuss unique proteins that mediate these events, the genes that encode them, and how they are regulated. A large number of articles focus on the areas of developmental immunology neurodevelopmental biology. The nervous system and the immune system, perhaps more than others, relies heavily on cell interactions (both cell-cell and cell-matrix) to function. Thus, neuronal tissue in different organisms and tissues involved in immune responsiveness serve as an excellent system to investigate the nature of the interactions needed for normal function.

The Contact A Site Glycoprotein Mediates Cell-Cell Adhesion by Homophilic Binding in *Dictyostelium discoideum*

C.-H. Siu, A. Cho, and A. H. C. Choi
J. Cell Biol., 105, 2523—2533, 1987 5-1

Contact sites A are EDTA-resistant binding sites expressed transiently on the cell surface of *Dictyostelium* cells during early development when cells become chemotaxic to cAMP. A membrane glycoprotein of 80,000 kDa (gp80) has been shown to play an important role in cell cohesion during this period, since antibodies to gp80 block the EDTA-resistant contact sites during the aggregation period. The present study was designed to investigate whether gp80 participates directly or indirectly in cell-cell binding, and if directly, how it does so.

gp80 was linked to fluorescent Covaspheres, and their binding to cells was assessed. Covaspheres bound strongly to developing cells with a stage specificity corresponding to that of cell surface appearance of gp80 and reaching a peak at 12 h. Cell aggregates bound to spheres much more strongly than did single cells. Control experiments demonstrated that Covaspheres linked to bovine serum albumin did not bind to cells and that Covaspheres linked to the lectin concanavalin A bound to cells, but not in a stage-specific manner. Trypsin treatment of Covaspheres abolished their binding to cells. When cells were pretreated with an adhesion-blocking anti-gp80 monoclonal antibody (80L5C4), binding of Covaspheres was inhibited. Covaspheres were capable of self-association, and this binding also was inhibited by pretreatment with 80L5C4. When midaggregation-stage cells were developed on polylysine-coated cover slips and then incubated with Covaspheres, scanning electron microscopy showed localization of the spheres at the two polar ends of the cells; this corresponded to the previously reported distribution of gp80 on the cell surface.

To measure the number of gp80 binding sites on aggregating cells, ^{125}I-labeled gp80 was bound to cells, permitting an estimate of 1.5×10^5 sites per cell; such binding was dose dependent, saturable, and blocked by pretreatment with 80L5C4. A nitrocellulose filter-binding assay showed dosage-dependent, specific binding of ^{125}I-labeled gp80 to immobilized gp80. Soluble gp80 blocked the reassociation of cells, again in a dosage-dependent fashion.

These experiments suggest that gp80 interacts homophilically *in vivo* and *in vitro*. The authors propose that gp80 has only one cell-binding site per molecule.

♦An 80,000-kDa glycoprotein (gp80) has been shown to be the major molecule of contact site A, which appears to be responsible for cell-cell cohesion in aggregating cells of *Dictyostelium*. Only aggregating cells could bind to the derivatized spheres. Soluble gp80 was also shown to abrogate the reassociation of disaggregated cells from the aggregation stage. The data suggest that the gp80 molecules interact in a homophilic manner. The relative involvement of protein and oligosaccharide moieties in this interaction require further study. *Stephen Alexander*

Surface Glycoprotein, gp24, Involved in Early Adhesion of *Dictyostelium discoideum*

D. A. Knecht, D. A. Fuller, and W. F. Loomis

Dev. Biol., 121, 277—283, 1987 5-2

In *Dictyostelium discoideum* development, three different cell-cell adhesion mechanisms appear sequentially. During the first 8 h, an EDTA-sensitive binding appears (contact site B). Later, EDTA-resistant binding arises (contact site A) which is inhibited by an antiserum to surface glycoprotein of 80,000 kDa gp80. A mutant in the post-translational glycosylation of gp80, *mod*B, does not display contact sites A, but still has early EDTA-sensitive adhesion properties, and a late EDTA-resistant adhesion mechanism that seems dependent on a 95,000-kDa glycoprotein. An earlier report from this laboratory described the production of antibodies that block early cell-cell adhesion (R695). The present paper describes the cell-surface antigen which neutralizes R695 and the developmental regulation of this 24,000-kDa glycoprotein (gp24).

To identify the antigen, solubilized membranes from *mod*B were fractionated, and the fractions assayed for their neutralization of the adhesion blocking activity of R695. All active fractions, and no inactive fractions, contained a 24,000-kDa protein, gp24. 4 μg of purified gp24 neutralized the activity of 18 μg of R695 IgG. A rabbit antiserum was produced against purified gp24 which was essentially monospecific on Western blots and as judged by immunoprecipitation. This IgG (R851) blocked early cell-cell adhesion only, while a combination of anti-gp24 and anti-gp80 blocked adhesion of 12-h cells; in *mod*B cells, R851 alone blocked all adhesion through 12 h. Purified gp24 completely neutralized the early adhesion blocking ability of R851.

Two monoclonal antibodies against gp24 were produced. One reacted with gp24 after the antigen was treated with pronase, but not after is was treated with periodate. The other monoclonal antibody gave opposite results. Apparently, the first recognizes a carbohydrate moiety, while the second recognizes a polypeptide portion of the molecule. Neither blocked adhesion or bound to cell surfaces. R851 may contain antibodies with the same specificity as these monoclonals, since the reactivity of this antiserum is partly inhibited by treatment of gp24 with both pronase and periodate monoclonal antibody showed that gp24 is synthesized only in developing cells, maximally during the first 4 h, but continuing at a lesser rate throughout development. Western blots demonstrated the presence of a constant level of gp24 in developing cells and the absence in growing cells. The coincidence of gp24 appearance with early EDTA-sensitive cell adhesion and the neutralization of adhesion blocking antibodies by purified gp24 provides evidence that gp24 is involved in contact site B.

♦*Dictyostelium* appears to have three distinct cohesion systems. The earliest of these is distinguished from the latter two by being sensitive to

EDTA. Antibodies were made to the membranes of cells harvested in the early stages of development. A glycoprotein of 24,000 Da was purified that was able to completely neutralized the cohesion blocking ability of the antibody. The gp24 antigen is absent in growing cells and begins to be synthesized at the onset of development. Synthesis is maximal during the first 4 h of development but continues at a lower rate throughout development. The results indicate that gp24 plays a role in the early cell-cell adhesion in this organism. *Stephen Alexander*

Post-Translational Glycosylation of the Contact Site A Protein of *Dictyostelium discoideum* is Important for Stability but not for its Function in Cell Adhesion

H.-P. Hohmann, S. Bozzaro, R. Merkl, E. Wallraff, M. Yoshida, U. Weinhart, and G. Gerisch

EMBO J., 6, 3663—3671, 1987 5-3

The contact site A protein of *Dictyostelium discoideum* (gp80) is a cell adhesion molecule which is modified by both cotranslational (type 1) and posttranslational (type 2) glycosylation. A previous report from this laboratory, which involved blocking type 1 synthesis with tunicamycin, suggested that type 1 carbohydrate was essential for neither the cell surface localization of gp80 nor for its stability, but did not rule out the possibility that it functions in cell adhesion. The experiments described here investigated *mod*B mutants, which are blocked in type 2 glycosylation.

In two independent mutant strains, a 68-kDa polypeptide was produced which was reactive to monoclonal antibodies to the gp80 polypeptide, but not to antibodies directed against type 2 carbohydrate. Both mutants produce less of this protein compared to the level of gp80 produced by wild-type cells. In both mutants, inhibition of protein synthesis with cycloheximide demonstrated a very short half-life for this 68-kDa protein compared to that of both gp80 and a related 66-kDa glycoprotein that lacks type 1 carbohydrate. In *mod*B cells, fragments of the protein were found in the medium during cell starvation, while in wild-type cells immunoreactive material remained intact on cell surfaces. Pulse-chase labeling experiments, which took advantage of the sulfation of the type I carbohydrate of the 68-kDa protein, suggested that the mutant protein is transported normally to the cell surface where it is then degraded, probably by cell-surface-bound proteases. Mutant cells displayed EDTA-stable adhesion during early development, although the aggregates formed were smaller than in wild-type cells.

These experiments suggest that the type 2 carbohydrate of contact site A protein is not required for cell-cell adhesion. Instead, they imply that the role of this moiety is in protecting the cell-surface exposed protein from proteolysis.

♦The contact site A glycoprotein (gp80) has two covalently linked oligosaccharides — an N-linked carbohydrate (type 1) and an apparently O-linked carbohydrate (type 2). Monoclonal antibodies to the polypeptide backbone of gp80 are able to detect, in *modB* mutants, the 68-kDa precursor of gp80, which has the type 1 but not type 2 sugars. In cells whose protein synthesis had been stopped using cycloheximide, the wild-type gp80 was stable over 4 h while the 68-kDa molecule, lacking the type 2 carbohydrate (*modB* mutants), rapidly decayed. Thus, the type 2 carbohydrate appears to confer stability on this molecule. Mutant cells (*modB*) acquired EDTA stable adhesiveness despite their lack of type 2 carbohydrate. The aggregates formed are smaller than wild type, presumably due to the lowered level of the protein at the cell surface. This study indicates that the type 2 carbohydrate on the contact site A molecule is not directly involved in cell-cell adhesion. *Stephen Alexander*

Gastrulation in the Sea Urchin Embryo Requires the Deposition of Crosslinked Collagen within the Extracellular Matrix

G. M. Wessel and D. R. McClay

Dev. Biol., 121, 149—165, 1987 5-4

During sea urchin morphogenesis, a dramatic increase in collagen synthesis and deposition in the extracellular matrix (ECM) coincides with gastrulation. The work presented here tested the hypothesis that this deposition is necessary for gastrulation by specifically disrupting collagen production and assaying developmental processes by morphological, biochemical, and immunological techniques. Three proline analogues were used in order to produce an unstable collagen of aberrant higher order structure. The effects of the lathyritic agent β-aminopropionitrile (BAPN) were also assessed. BAPN inhibits extracellular collagen cross-linking and results in collagen molecules lacking the stability or strength to function normally.

When embryos were exposed to these agents, they developed normally until an arrest occurred at the mesenchyme blastula stage, and they failed to gastrulate, spiculate, or produce pigment cells. When the agents were removed, normal development resumed. BAPN did not appear to inhibit either respiration or protein synthesis at any of several stages tested. Failure to accumulate collagen in the ECM was shown to coincide with failure to gastulate, and when BAPN or proline analogues were removed, recovery of collagen synthesis coincided with return to normal development. The effects of BAPN treatment on the expression of 15 different developmentally expressed molecules were assessed. In 14 cases, no effect was detected, but the synthesis of an endoderm specific molecule, Endo 1, normally expressed during gastrulation, was reversibly inhibited.

These experiments reveal a correlation between collagen deposition in the ECM and gastrulation. The techniques employed in this work appear to affect the collagenous matrix specifically and reversibly, and should thus prove useful for further study on the effects of cell-ECM interaction.

♦The extracellular matrix has long been implicated to have important roles for morphogenesis in the sea urchin embryo (Gustafson and Wolpert, 42, 442—498, 1967). Current efforts in this field are in identifying what molecules are present, when in development they appear, and what possible function they might perform. It appears that the sea urchin contains components immunologically similar to the extracellular matrix components found in vertebrates, and several laboratories have focused on the possible role of collagen in development. In the sea urchin embryo, collagen has been identified by ultrastructure, by amino acid composition, by associated processing enzymes, and recently by nucleic acid sequence (all referenced in Angerer et al., *Genes Dev.*, 2, 239, 1988). Here, Wessel and McClay establish a correlation between collagen accumulation in the extracellular matrix and certain events of gastrulation and spiculogenesis. Disrupting collagen processing by proline analog incorporation or by a lathrytic agent inhibit gastrulation and spiculogenesis. However, the embryos are able to recover from this treatment, reaccumulate collagen, and proceed with gastrulation. These findings may be useful for examining the role of collagen in gastrulation, studying cell-extracellular matrix interactions, and in exploring the mechanisms of gastrulation. *Gary M. Wessel*

Regulation of Extracellular Matrix Assembly: In Vitro Reconstitution of a Partial Fertilization Envelope from Isolated Components

P. J. Weidman and B. M. Shapiro
J. Cell Biol., 105, 561—567, 1987 5-5

Fertilization of sea urchin eggs is followed by the transformation of the glycocalyx, or vitelline layer, into a prominent fertilization envelope. In an attempt to understand the assembly and modification of the fertilization envelope, this laboratory has previously reported the purification of ovoperoxidase and proteoliaisin, two components of the fertilization envelope of sea urchin eggs. Here, the binding of these purified, ^{125}I-labeled proteins to sea urchin eggs was measured as a test of the hypothesis that proteoliaisin permits the union of ovoperoxidase with vitelline layer components.

Ovoperoxidase binding to unfertilized eggs was completely and linearly dependent on the presence of proteoliaisin. This binding required Ca^{2+}, as does the interaction of these two proteins. Proteoliaisin, however, bound to eggs in the presence of only divalent cations, and the binding was maximum in seawater containing both Ca^{2+} and Mg^{2+}. Saturation binding curves for proteoliaisin binding to unfertilized eggs indicated that only one class of binding sites exist. Dissociation constants and number of binding sites on each egg were determined, but these values varied with the divalent cation composition of the artificial seawater medium used. The interaction of proteoliaisin with the vitelline layer was unaffected by a variety of monosaccharides or complex carbohydrates.

These experiments imply that proteoliaisin binds to the vitelline layer directly, and thereby mediates the subsequent binding of ovoperoxidase. This work also resulted in achieving the *in vitro* assembly of a partial fertilization envelope from purified components.

♦Weidman and Shapiro have explored a type of self-assembly in the extracellular matrix of the sea urchin egg. This matrix, termed the vitelline layer, is modified during the fertilization reaction to form a protective envelope important during early embryogenesis. Many of the modifications to the vitelline layer occur by the addition of specific proteins from the egg cortical granules, resulting in a precise structure then called the fertilization envelope. This report addresses the mechanisms of these modifications *in vitro*. The model resulting from this study demonstrates the importance of specific binding of the protein, proteoliaisin, to the vitelline layer. This noncovalent interaction is divalent cation (Ca^{++} and Mg^{++}) mediated. The proteoliaisin-vitelline layer interaction then permits the addition of the second component, ovoperoxidase, into a specific conformation on the vitelline layer. The function of the ovoperoxidase is to covalently couple proteins within the fertilization envelope, via a tyrosine cross-linking reaction. Thus, the proteoliaisin-mediated orientation of ovoperoxidase appears crucial for the directed, covalent cross-linking necessary for this specific extracellular structure. These results are very useful because for one, they demonstrate the importance of molecular self assembly in the extracellular matrix. In addition, defining this *in vitro* system should be quite useful in studies to determine the structural properties required for these important protein interactions. *Gary M. Wessel*

Cell-Cell Interactions in the Guidance of Late-Developing Neurons in *Caenorhabditis elegans*

W. W. Walthall and M. Chalfie

Science, 239, 643—645, 1988 5-6

The *Caenorhabditis elegans* touch receptor, AVM, is a late-developing neuron, which must navigate through the mature neuropil of the nerve ring to make appropriate contacts. The effect of cell-cell interactions on its growth were assessed by laser ablation of a pair of interneurons, the right and left BDU cells, that receive synapses from AVM.

In animals in which the BDU cells were destroyed at hatching, AVM could not function in the touch response. Therefore, BDU cells are needed for the AVM touch response. Genetic ablation with an *unc*-86 mutation produced the same result. However, if the BDU cells were killed 24 h after hatching, the touch response was not affected. Therefore, the BDU cells do not appear to be part of the touch-reflex circuit itself.

The BDU cells are necessary to the touch-reflex response only during the stage when the AVM cell is growing toward its target. Therefore, it appears likely that the BDU cells guide the AVM branch to its proper synaptic site. This form of guidance may be more important when late-developing processes must grow through an established nervous system.

♦This paper represents the beginning of a detailed analysis of cell-cell interactions that are important to nerve development in *C. elegans*. The real advantage of identifying interactions with cells, such as BDU, that are required for pathfinding by other neurons is that once these interactions are revealed, one can begin to identify mutations in genes that are involved in the interactions. In this way we can hope to "understand" at a molecular level, how a nervous system is put together. *Joseph G. Culotti*

Reversal of Cellular Polarity and Early Cell-Cell Interaction in the Embryo of *Caenorhabditis elegans*

E. Schierenberg

Dev. Biol., 122, 452—463, 1987 5-7

During *Caenorhabditis elegans* early embryogenesis, the first cell division generates a larger anterior somatic cell and a smaller posterior germ-line cell, P1. P1 then cleaves unequally into a larger somatic cell, EMS, and the germ-line cell P2. P2 then divides into a larger somatic, E, and a smaller germ-line cell, P3. The role of anterior-posterior cleavage polarity during these cell divisions was investigated by puncturing the eggshell with a laser microbeam and extruding parts of the embryo.

After the divisions of P0 and P1, the somatic cell was anterior to the germ-line cell. After the divisions of P2 and P3, the germ-line cell was anterior to the somatic cell (see Figure 2, *Dev. Biol.*, 122, 454, 1987 and Figure 5-7). Observation of extruded egg fragments and of the partial twins that result indicated that if the resulting mirror image duplications were derived from P0, they were joined at the somatic cell, while if they were derived from P1, they were joined at the germ-line cell. Therefore, it appears that the anterior-posterior polarity of the germ line of *C. elegans* is reversed in the P2 cell as a part of normal development. When the P2 cell was extruded from a 4-cell embryo, the EMS cell underwent its normal cleavage. However, its daughter cell, E, was not positioned normally and divided aberrantly. Therefore, an interaction between P2 and EMS appears to be necessary for the normal unequal cleavage of EMS.

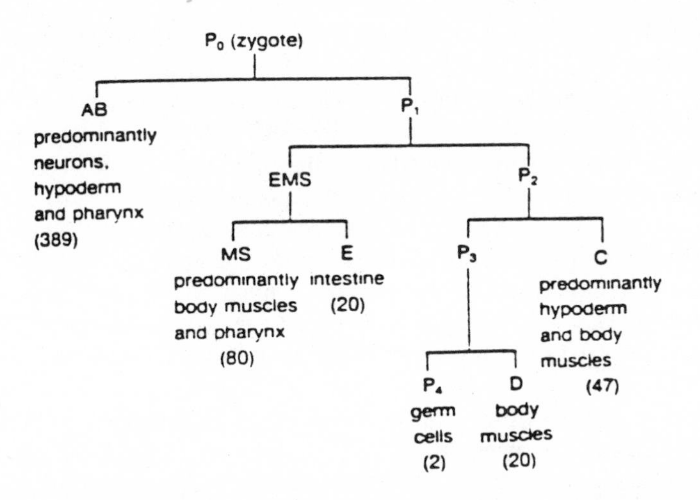

FIGURE 5-7. Modified early lineage tree showing the five somatic founder cells (AB, MS, E. C, D) and the germ line (P_0-P_4). Following the notation rules that a cell anterior relative to its sister is placed on the left arm of a lineage branch (Deppe et al., 1978) P_3 and C and P_4 and D, respectively, have exchanged positions compared to previous versions of the lineage tree. (From *Dev. Biol.*, 122, 455, 1987. With permission.)

This paper reports on the occurrence of a polarity reversal within a single cell (P2) that occurs as part of the normal development of *C. elegans.* It is possible that this polarity reversal allows P2 to interact with EMS and to undergo its normal unequal cell division.

♦This paper provides additional evidence for the use of cell-cell interactions during early embryogenesis to determine cell division patterns and cell fates in *C. elegans*. This is a fascinating paper. The data strongly suggest that the potential for germline-like cleavage (asymmetric) resides in the posterior of the P0, P1, and P2 germline precursors, but moves to the anterior of P2 as its cell cycle progresses. This may account for the reversal in polarity of asymmetric germline divisions seen in P2. Other evidence suggests that P2 and EMS interact and this interaction effects the asymmetric positions of both nuclei as well as the asymmetric division of EMS. *Joseph G. Culotti*

Sevenless, a Cell-Specific Homeotic Gene of *Drosophila*, Encodes a Putative Transmembrane Receptor with a Tyrosine Kinase Domain

E. Hafen, K. Basler, J.-E. Edstroem, and G. M. Rubin
Science, 236, 55—63, 1987 5-8

The *sevenless* gene of *Drosophila* is responsible for the development of the R7 photoreceptor cell. This paper reports the cloning and characterization of the *sevenless* gene.

The polytene chromosome microdissection technique was used to obtain DNA from the *sevenless* gene region. This was used to obtain overlapping clones from genomic DNA libraries. The position of these clones was determined by *in situ* hybridization. P-element mediated germ-line transformation was employed to ensure that the entire gene had been cloned. Expression of the *sevenless* gene was detected in heads, but not bodies of adult flies. In third instar larvae, *sevenless* expression was detected only in the posterior two thirds of the eye disk. The longest cDNA that was obtained was sequenced and was missing some 5′ sequences. Conceptual translation of this sequence indicated that the product of this gene would be larger than 220 kDa. A transmembrane stretch of 22 amino acids divided the protein into an extracellular domain of 172 kDa and an intracellular domain of 45 kDa. In the carboxy-terminal region a significant homology to tyrosine kinases was detected.

On the basis of the sequence of the *sevenless* gene, it was hypothesized that the *sev* protein directly reads positional information specifying the R7 developmental pathway via its extracellular domain. This signal would be transmitted to the cell through the tyrosine kinase activity of the carboxy-terminal protein.

Molecular Characterization and Expression of *sevenless*, a Gene Involved in Neuronal Pattern Formation in the *Drosophila* Eye

U. Banerjee, P. J. Renfranz, J. A. Pollock, and S. Benzer

Cell, 49, 281—291, 1987 5-9

The *Drosophila sevenless* mutation causes flies to lack the photoreceptor neuron R7. This paper describes the cloning of this gene, its expression, and the identification of its product.

New *sev* alleles were created by hybrid dysgenesis. One of these alleles, *sevP1*, was further characterized. This mutant gene was cloned and used to isolate overlapping genomic clones from wild-type libraries. These were used to isolate nine cDNA clones. The longest clone, c2(3), was 4.7 kb. This clone recognized an 8.2-kb transcript which was expressed in late third instar imaginal disks, 5- to 12-h whole pupae, and adult heads. *In situ* hybridization showed *sev* transcripts on the apical surface from just behind the morphogenetic furrow to the posterior of the eye disk in late third instar larvae. Antibody to the *sev* product showed specific staining in some of the cells in the same region of the disk.

These experiments have demonstrated that the *sev* gene makes a long transcript, which is expressed posterior to the morphogenetic furrow in late third instar eye imaginal disks. The antibody staining indicated that *sev* product is expressed in a subset of cells within each photoreceptor cluster.

Localization of the *sevenless* Protein, a Putative Receptor for Positional Information, in the Eye Imaginal Disc of *Drosophila*

A. Tomlinson, D. D. L. Bowtell, E. Hafen, and G. M. Rubin
Cell, 51, 143—150, 1987 5-10

It has been proposed that the *sevenless* gene product receives positional information required for the development of the R7 photoreceptor. Antibodies to the *sev* protein were generated, and its expression was examined at both the light and electron microscope levels.

Light microscopic studies indicated that staining with the *sev* antibody begins at the morphogenetic furrow and extends posteriorly. To identify the specific cells stained by this antibody, immunocytochemistry was performed and examined by electron microscopy. The *sev* protein was first detected at the precluster stage (see Figure 2, *Cell,* 51, 146, 1987) in R3, R4, and in two cells of unknown fate. By the symmetrical cluster stage, R7 and the anterior and posterior cone cells stained heavily. At the 2-cone cell stage, the polar cone cell began to stain, staining in R7 was reduced, and the equatorial cone cell darkened. At the 4-cone cell stage, staining in the anterior and posterior cone cells was reduced. This staining pattern is summarized in Figure 5 (see *Cell,* 51, 148, 1987). The heaviest staining region of all these cells was the microvilli. Staining was also detected in large membrane bound vesicle-like structures. Plasma membrane staining occurred at adherens-type junctions that face R8.

The product of the *sevenless* gene is detected prior to the specification of R7. Therefore, *sev* may be involved in receiving signals involved in the specification of R7 development. However, as *sev* is active in cells, such as R3 and R4, that are not involved in this pathway, it is not likely that *sev* alone is sufficient to specify this developmental pathway.

The *sevenless+* Protein is Expressed Apically in Cell Membranes of Developing *Drosophila* Retina; It Is Not Restricted to Cell R7

U. Banerjee, P. J. Renfranz, D. R. Hinton, B. A. Rabin, and S. Benzer
Cell, 51, 151—158, 1987 5-11

When the *sev* gene of *Drosophila* is mutant, photoreceptor R7 does not develop. A monoclonal antibody was prepared against the product of the *sev* gene. This antibody was used to determine the ultrastructural location of the *sev* protein in the third instar larval eye disk.

Specific staining was detected immediately behind the morphogenetic furrow of the eye disk. In more posterior regions of the disk, the staining was localized over each developing photoreceptor cluster. The *sev* protein was located apically, limited to approximately 1 μm from the epithelial surface. Electron microscopy was employed to examine expression at the cellular level. Staining was localized to the plasma membranes of microvilli that form the tuft at the apex of each photoreceptor cluster and to apical cell membranes. Staining was not restricted to cell R7, but occurred in all eight photoreceptor cells and in the four peripheral cone cells.

A monoclonal antibody to the *sev* gene product has been used to localize this protein within the eye disk. Surprisingly, *sev* expression is not limited to photoreceptor R7, but occurs in all the photoreceptors, as well as in cone cells. The protein is abundant in apical tip membranes and microvilli, which is consistent with its role as a receptor.

Nucleotide Sequence and Structure of the *sevenless* Gene of *Drosophila melanogaster*

D. D. L. Bowtell, M. A. Simon, and G. M. Rubin

Genes Dev., 2, 620—634, 1988 5-12

The R7 photoreceptor does not form in *Drosophila* carrying the *sevenless* mutant. This paper describes the cloning, sequencing, and characterization of the *sev* gene.

Overlapping cDNA clones were isolated from an eye imaginal disk library. These clones were used to isolate corresponding clones from a genomic library. Clones were sequenced and transcription start sites were mapped (Figure 5-12). Conceptual translation of the cDNA sequence demonstrated an open reading frame of 7677 nucleotides, which corresponds to 2559 amino acids. This protein is predicted to be 288 kDa, which is larger than is typical for tyrosine kinase receptors. An amino-terminal hydrophobic sequence is suggestive of a transmembrane anchor, but could protentially be cleaved. If the amino-terminal region is an anchor, there would be a large extracellular loop, attached both at the amino terminus and at the transmembrane domain. The transmembrane domain is adjacent to the cytoplasmic carboxy-terminal kinase domain.

This paper describes the structure of the *sevenless* gene and its predicted product. This product differs from other tyrosine kinase receptors in the large size of its extracellular domain and the possible presence of two transmembrane domains. These features may be important in the function of this protein as a receptor for positional information.

♦Mutants in the *sevenless* (*sev*) locus of *Drosophila* have an extremely interesting phenotype in that they affect the developmental fate of only one (R7) of the 20 cells that make up the ommatidium, the functional unit of the adult eye. The compound *Drosophila* eye consists of a two-dimensional array of about 800 ommatidia. Each ommatidium contains 20 cells: 8 photoreceptor neurons whose axons pass through the optic stalk to communicate with the brain, 4 cone cells that build the liquid-filled lumen underlying the lens, 1 sensory hair and nerve, and 3 types of pigment cells that optically insulate each ommatidium. In *sev* mutants the cell that would normally develop into an R7 photoreceptor neuron develops into a non-neuronal cone cell. The developmental program of all other cells is normal.

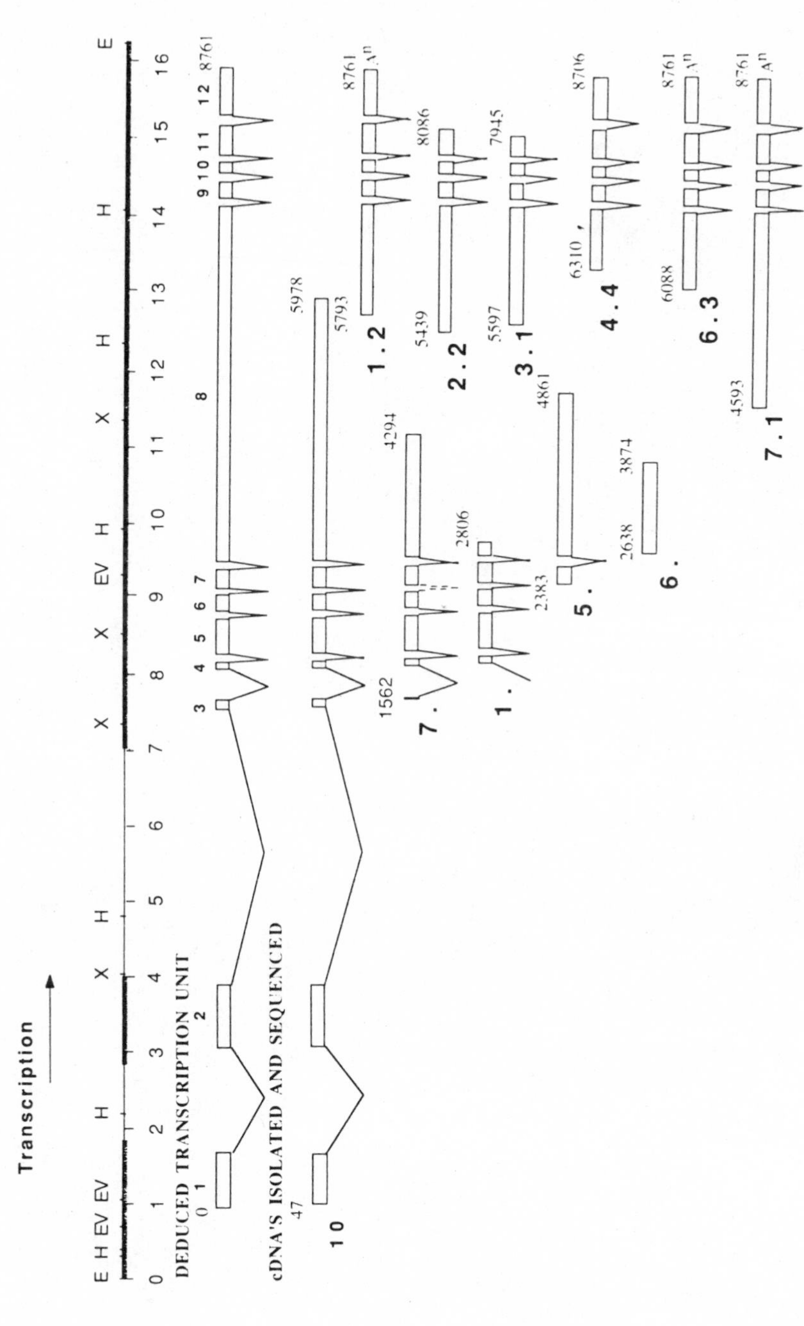
Transcription
DEDUCED TRANSCRIPTION UNIT
cDNA'S ISOLATED AND SEQUENCED
E H EV EV EV H X H EV X EV H X H X H H E
0 1 2 3 4 5 6 7 8 9 10 11 12 13 14 15 16
1 2 3 4 5 6 7 8 9 10 11 12
1.2
2.2
3.1
4.4
6.3
7.1
7.
1.
5.
6.
10
An

FIGURE 5-12A

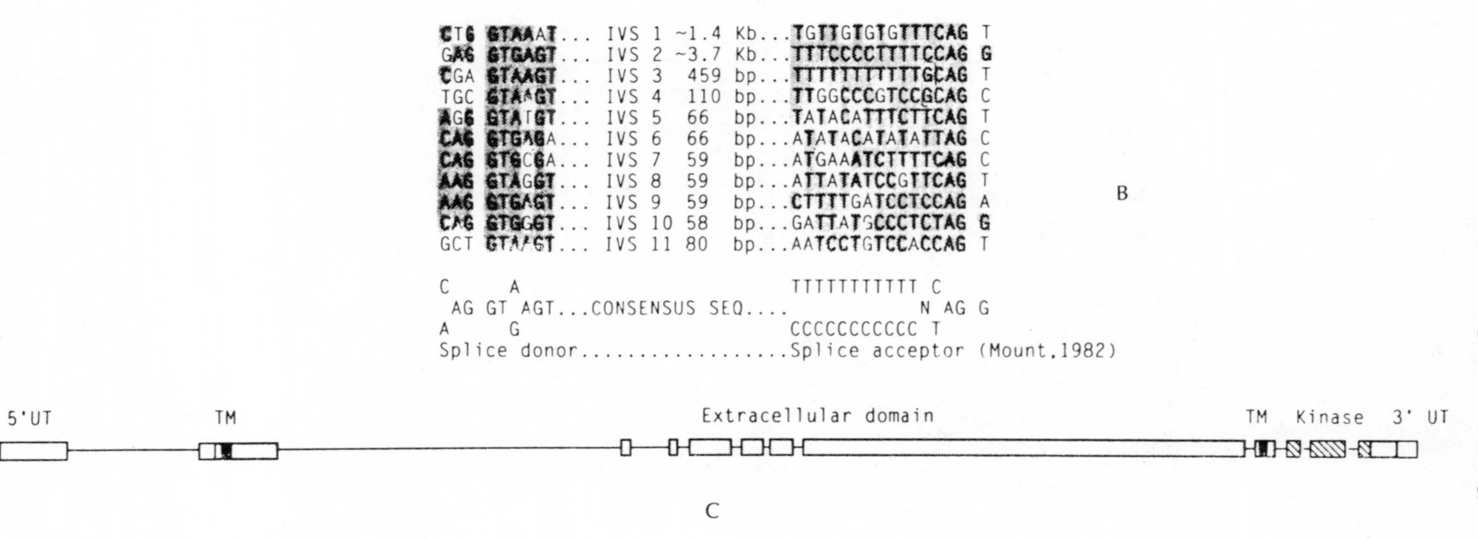

FIGURE 5-12. Intron-exon structure of the *sevenless* gene deduced form the sequence of 11 cDNA fragments and most of the 16.3-kb genomic *Eco*RI fragment. (A) Restriction map of the genomic fragment used to rescue the *sevenless* phenotype (Hafen et al. 1987) (E, *Eco*RI; Ev, *Eco*RV; H, *Hind*III; X, *Xho*I). Bold lines indicate the portion of this fragment sequenced. Below is the deduced structure of the transcription unit and the 11 cDNA fragments from which it was derived. Numbering of these fragments refers to the cDNA sequence shown in Figure 2 (see *Genes Dev.*, 2, 624, 1988), beginning at the major transcriptional start site and ending at the polyadenylation site (nucleotide 8760). Numbers in boldface type identify each clone. Clones 7 and 1 are partial exceptions to the deduced structure; clone 1 includes 54-bp of intron 3, and clone 7 retains intron 6. However, these clones probably represent partially processed RNA precursors. (B) Genomic sequences surrounding the splice acceptor and donor sites and the length of each intron. Bases conforming to the consensus sequences (Mount 1982) for these sites are shaded. (C) Schematic representation of the various domains of the *sevenless* protein. (TM) Putative amino-terminal and carboxy-terminal transmembrane domains; (UT) 5′- and 3′-untranslated regions; (kinase) putative tyrosine kinase domain. (From *Genes Dev.*, 2, 620, 1988. With permission.)

Two properties of eye development are important for the present studies. Unlike certain other invertebrates, there is no clonal lineage in the development of the *Drosophila* eye; the developmental fate of a particular cell is unrelated to its ancestors and presumably depends only on the position it occupies relative to other cells in the field (cf. cell-cell communication). The *sev* mutation is completely autonomous. For instance, in genetic mosaics a single prospective *sev*$^-$ R7 will develop into a cone cell, even if completely surrounded by *sev*$^+$ cells. Conversely, a prospective *sev*$^+$ R7 develops normally, even if completely surrounded by *sev*$^-$ cells. The simplest hypothesis to explain these properties is that the *sev* gene is expressed only in R7 and that its function is to direct the cell towards the correct developmental pathway. As reviewed below, the experimental data did *not* support this model.

These papers report on the cloning and sequencing of the *sev* gene, the generation of antibodies against the *sev* protein, and the localization of the protein at the electron microscope level. The nucleotide sequence of *sev* predicts a large, 288-kDa protein which contains from the amino terminus, a short cytoplasmic domain followed by a hydrophobic, transmembrane amino acid stretch followed by a large extracellular loop of over 2000 amino acids and a second transmembrane stretch that "reinserts" the protein into the cytoplasm and finally, at the carboxy terminus, a cytoplasmic region with homology to the tyrosine kinase domain of hormone receptors. The *sev* gene is expressed in the developing eye several hours before overt differentiation of R7. Surprisingly, immunoelectron microscopy revealed that expression of *sev* is not restricted to the R7 precursor. It is expressed according to a complex temporal and spacial pattern in many of the cells that will form the ommatidium, including the non-neuronal cone cells. While the bulk of the protein appears to be located in the apical microvilli facing the extracellular matrix, staining was also observed at junctional complexes connecting specific precursor cells. These observations do not lead to a simple model for the *sev* mechanism of action. Further experiments are required to generate additional clues.

In conclusion, much progress has been made in the understanding of this interesting gene. Despite the simple phenotype of mutants, it is not yet possible to encompass all the data in a simple model. Given the absence of a defined cell lineage, cell-cell communication is an essential ingredient for any hypothesis and the localization of the protein in membranes satisfies this requirement. It is likely that additional genes are involved in the transmission of the signal(s), and mutations in these genes may yield phenotypes that are similar or related to that of *sev*. *Marcelo Jacobs-Lorena*

Cell Autonomy of Expression of Neurogenic Genes of *Drosophila melanogaster*

G. M. Technau and J. A. Campos-Ortega
Proc. Natl. Acad. Sci. U.S.A., 84, 4500—4504, 1987 5-13

Fruit flies that are defective in neurogenic genes are deficient in epidermal cells and have excess neuronal cells. To test the cell autonomy of the neurogenic phenotype, individual cells were transplanted from neurogenic ectoderm of embryos mutant for *almondex, big brain, master mind, neuralized, Delta,* and *Enhancer of split* to wild-type embryos.

The transplanted cells from these mutants gave rise to clones with a normal distribution of epidermal and neuronal histotypes, except for *Enhancer of split. E(spl)* mutant cells gave rise exclusively to neuronal cells. Therefore, the expression of *Enhancer of split* is autonomous, while all other tested neurogenic genes are not autonomous.

This is consistent with a model in which the nonautonomous genes provide a regulatory signal, which was supplied by surrounding wild-type cells in these experiments. The autonomous gene would encode a receptor for this signal. Receipt of the signal would be required for commitment to a neuronal developmental pathway.

♦During early *Drosophila* embryogenesis, cells of the neurogenic ectoderm become committed to differentiate along either the epidermal or neurogenic pathways. Several lines of evidence indicate that cell-cell interaction plays an important role in this commitment process, implying the existence of a signal emitted by one cell and the reception of this signal by another cell. In embryos carrying a loss-of-function mutant of neurogenic genes most ectodermal cells develop into neurons. To test the autonomy of the mutant neurogenic phenotype, single marked cells were transplanted from mutant donor embryos into wild type host embryos. Should the mutation be autonomous, injected mutant cells would be expected to develop almost exclusively into neural tissues. Only one, *enhancer of split,* out of seven mutants tested behaved in this way. Cells of the remaining six neurogenic mutants (*almondex, big brain, master mind, neuralized,* and *Delta*) developed into both neural and epidermal tissues at a frequency similar to injected wild-type cells. The interpretation of this result is that the mutant phenotype of the injected cell was "corrected" by the surrounding wild-type host tissue. The model suggested by these data is that mutants in this class have a functional "receptor" but are defective in sending a hypothetical intercellular regulatory signal; this signal would be provided by the surrounding wild type cells. On the other hand, *enhancer of split* because of its autonomy even at the single cell level (see below), may correspond to a "receptor"-coding gene. Interestingly, previous experiments had been done where larger patches of mutant cells (rather than the single cell injection of this work) were generated by mitotic recombination. In this case, expression of the mutant phenotype was autonomous rather than nonautonomous as in the present work. This may mean that the signal cannot travel over large distances (many cell diameters) and may depend on direct cell-cell interaction.

These elegant experiments provide some important clues concerning the mechanism of action of genes that influence differentiation along the epidermal and neurogenic pathways. The reader is also referred to report 11 in this volume for comments on related work. *Marcelo Jacobs-Lorena*

The *Drosophila* Clock Gene *per* Affects Intercellular Junctional Communication

T. A. Bargiello, L. Saez, M. K. Baylies, G. Gasic, M. W. Young, and D. C. Spray

Nature (London), 238, 686—691, 1987 5-14

Changes in the *per* gene of *Drosophila* can alter circadian and behavioral rhythms. In an effort to understand the function of the *per* gene product, *per* mRNA and protein were localized in embryos and larvae.

In situ hybridization was performed on 10- to 20-h embryos. In all embryos older than 10 h, labeling was seen exclusively in the salivary gland. Northern analysis indicated that the 4.5-kb *per* transcript can be detected in 8- to 12-h embryos and that peak accumulation occurred in 12- to 16-h embryos. Embryonic salivary glands were also labeled by antibody directed against the *per* protein. This protein was detected at cell boundaries and at the cell surface. Dye coupling experiments were used to examine cell communication in wild-type and *per* mutant flies. In wild-type flies Lucifer Yellow readily diffused between cells of the salivary gland. In *per0* flies there was no diffusion. In *perS* flies, the transfer of dye between cells of the salivary gland was more rapid than in wild type. The strength of electrotonic coupling between salivary gland cells was also examined. There was very weak coupling in *per0* and *per–* glands, while in *perS* glands the coupling was greater than in wild-type glands.

These data are consistent with a model in which the product of the *per* locus is involved in intercellular junctional communication in *Drosophila*. Therefore, the *per* locus is probably not the biological clock itself. However, changes in intercellular junctional communication could affect the nervous system biological clock.

An Inducible Promoter Fused to the *period* Gene in *Drosophila* Conditionally Rescues Adult *per*-Mutant Arrhythmicity

J. Ewer, M. Rosbash, and J. C. Hall

Nature (London), 333, 82—84, 1988 5-15

The *Drosophila per* gene is involved in the expression of circadian rhythms. P element-mediated transformation was used to create conditional *per* mutants. Arrhythmic *per0* flies were transformed with a wild-type *per* gene linked to the *hsp70* promoter. The activity pattern of these flies was assessed at high and low temperature.

Rhythmicity could be induced in three lines of transformed flies by increasing the temperature from 18 to 29° C, which induced the *hsp70* promoter. These flies were transferred from 18 to 29° C at various stages of the life cycle to determine when *per* expression is required. *Per* expression in the adult was both necessary and sufficient for rhythmic behavior. The transfer of these adult flies to 29° C at any point rapidly induced a rhythmic phenotype. Therefore, *per* did not appear to be required for light/dark entrainment. Without the continued presence of the *per* product, rhythmic behavior was rapidly lost.

Therefore, *per* mutations do not appear to disrupt the biological clock of *Drosophila*. The *per* gene appears to be required in the adult fly to manifest the locomotor rhythmicity of an underlying circadian oscillator.

Spatial and Temporal Expression of the *period* Gene in *Drosophila melanogaster*

X. Liu, L. Lorenz, Q. Yu, J. C. Hall, and M. Rosbash

Genes Dev., 2, 228—238, 1988 5-16

The temporal and spatial expression of the *Drosophila per* gene was examined by *in situ* hybridization and by staining of flies transformed with a *per*-beta-galactosidase fusion gene.

The *in situ* and chromogenic (X-gal) results were always identical. Staining of fusion embryos with X-gal was observed at stage 14 (11 h) in ganglia. Beta-galactosidase staining was not observed in larvae. Staining was again observed in mid-to-late pupae in the ring gland and optic lobes. In the adult, head, eyes, optic lobes, antennae, ocelli, brain, proboscis, and esophagus stained positively for beta-galactosidase. Halteres, legs, and thoracic and abdominal ganglia also stained. Many tissues of the abdomen were also positive for *per* expression.

Since so many tissues stain positively for *per* gene expression, it is possible that each contains an intrinsic circadian oscillator.

♦Much progress has recently been made in the study of the *Drosophila period* (*per*) gene. Mutant alleles exist that lengthen, shorten, or abolish the periodicity of the circadian rhythms. The same alleles also alter the frequency of the wing beat during the male courtship song. Mutations in the *per* locus also have other effects, such as on the frequency of the larval heartbeat and on the uptake by isolated salivary glands of a voltage-sensitive dye. The gene has been cloned and preliminary evidence suggests that the *per* protein may be a proteoglycan.

The starting point of Bargiello et al.'s studies is their observation that *per* antibodies strongly stain embryonic and larval salivary glands around their cell periphery (suggesting association with cell membranes), and the observation by others that dye uptake in isolated salivary glands is influenced by circadian rhythms. Bargiello et al. measured cell communication in salivary glands of wild type and *per* mutant flies by dye transfer and by electrical coupling experiments. Coupling was very weak in *per* null mutants, was high in wild type flies, and even higher in *per*^{s} mutants, which exhibit an even shorter circadian period than wild type. This report provides strong evidence that the *per* gene product influences cell-cell communication in salivary glands and raises the interesting possibility that this may also happen in the nervous tissue. Thus, the *per* gene product is certainly an essential component in the pathway that leads from the biological clock to biological rhythms. It is likely that it acts as a transducer of the signal rather than being the source of the clock itself. The observations by Liu et al. that many tissues of the adult express the gene and by Ewer et al. that the rhythms are only dependent on the immediate past history of *per* expression (see below) are consistent with this interpretation.

Liu et al. linked the beta-galactosidase reporter gene to the *per* gene promoter and transformed this construct into the *Drosophila* germ line. By staining for beta-galactosidase, they established in which tissues the *per* gene promoter acted. In general, there was good agreement with the previous localization of the *per* mRNA by *in situ* hybridization. The fusion gene was active in the midline nervous tissue of mid-to-late embryos and in the prothoracic gland-corpora allata and the optic lobes of the pupae. Surprisingly, a large number of tissues stained in the adult.

Ewer et al. constructed a *per* gene driven by the strong and inducible heat shock promoter and introduced it into *per* null flies. They then determined the relationship between time of *per* expression and restoration of circadian rhythms. They found that rhythmic behavior is independent of the history of *per* expression earlier in development. Moreover, *per* product is not required for the "learning" part: light/dark entrainment in the absence of *per* product is entirely effective, the *per* product only being required for the execution of the rhythmic behavior. This is an important set of experiments that further sheds light on the mechanism of *per* action. These results are consistent with the hypothesis that *per* is not the pacemaker itself but only serves to transduce the pacemakers signals.

Two earlier papers (Yu et al., *Proc. Natl. Acad. Sci. U.S.A.*, 84, 784—788, 1987; and Yu et al., *Nature (London)*, 326, 765—769, 1987) that were published prior to the period covered by this survey, identify at the nucleotide level lesions in *per* mutant genes. *Marcelo Jacobs-Lorena*

Reinvestigation of the Role of the Optic Vesicle in Embryonic Lens Induction

R. M. Grainger, J. J. Henry, and R. A. Henderson
Development, 102, 517—526, 1988 5-17

Induction of the lens by the optic vesicle in amphibians often is used to support the view that a single inductive event can be determining in a multipotent tissue. Transplantation studies have indicated that many regions of embryonic ectoderm that normally would form epidermis can induce a lens if brought into contact with the optic vesicle. The authors conducted optic vesicle transplants in *Xenopus* embryos and also in *Rana palustris,* the organism first used in these studies.

Contamination of optic vesicles by presumptive lens ectoderm cells can generate lenses in *Xenopus laevis*. When horseradish peroxidase was used as a lineage tracer, the optic vesicle was unable to stimulate lens formation in neurula- or gastrula-stage ectoderm. In careful studies with *Rana palustris*, the optic vesicle alone did not induce lens formation. The transplanted tissues were placed some distance from the head ectoderm, beneath the lateral or ventral ectoderm.

Induction of several tissues other than lens appears to involve multiple steps; the nose and ear are examples. Mesodermal induction and neural induction often are stated to be single-step processes, but in both instances the events are quite complex. Hopefully, future studies will show whether multiple steps are required for induction of these tissues, as appears to be the case for the lens.

♦This paper refutes the famous classical result of Spemann that the optic cup can induce a lens in amphibian ventral ectoderm. Grainger and colleagues lineage-labeled donors and recipients of grafts with horseradish peroxidase and showed convincingly that the lenses that form in mid-neurula belly ectoderm come from the transplanted cup rather than the host. They also show that at the midgastrula stage presumptive head ectoderm can be induced by a donor cup, and that it is much more receptive to this induction than ventral ectoderm of the same stage. These results fit nicely with the models proposed by Anton Jacobson 20 years ago (Jacobson, *Science,* 152, 25—34, 1966), in which lens and other sensorial organs are induced in a multistep sequence. *Thomas D. Sargent*

Heparan Sulfate Proteoglycan and Laminin Mediate Two Different Types of Neurite Outgrowth

D. Hantaz-Ambroise, M. Vigny, and J. Koenig

J. Neurosci., 7, 2293—2304, 1987 5-18

When spinal cord neurons are cultured *in vitro*, they respond to environmental changes by both neurite elongation and the emergence and branching of new neurites. Neurite outgrowth appears to be regulated by at least two components of the basal lamina, heparan sulfate proteoglycans (HSPG) and laminin. The authors cultured rat embryo spinal cord neurons with HSPG medium and with laminin-supplemented culture medium, as well as with conditioned medium from primary embryonic rat muscle cultures (MCM). Neurites were analyzed by the binding of tetanus toxin antitoxin fluorescein-bound antibody conjugates.

Culture of cord neurons with HSPG medium led to a threefold increase in neurite elongation. Extensive branching occurred on culture of neurons with laminin-supplemented medium. These responses were blocked by specific anti-HSPG and anti-laminin sera, respectively. Conditioned medium from MCM cultures mimicked the effects of both HSPG and laminin. Immunoprecipitation with anti-HSPG and anti-laminin antibodies showed that MCM contains the two basal lamina components.

Laminin and HSPG affect neurite outgrowth from spinal cord neurons in different ways. Elongation and terminal branching are observed in reinnervation and after experimental muscle paralysis. The neurite responses might correlate with differences in the nature or level of basal lamina components, as well as with trophic factors released by inactive muscle.

♦In its native state, laminin forms complexes with other molecules such as entactin/nidogen and heparan sulfate proteoglycan. These molecules may contribute to the functional activity of laminin by changing its configuration or by themselves promoting cell attachment. The neurite-promoting region of laminin is on the long arm of the laminin molecule in the region of the heparin binding site. In fact, antibodies against the heparin binding site affect neurite outgrowth on laminin. However, as described above, this may be due to steric hindrance. Another possibility is that both laminin and heparan sulfate proteoglycans can affect neurite outgrowth independently.

This paper by Hantaz-Ambroise et al. explores the respective roles of laminin and heparan sulfate proteoglycan on spinal cord neurons in culture. The most pertinent finding is the *both* laminin and heparan sulfate proteoglycan promote neurite outgrowth, though the pattern of outgrowth is somewhat different since the former promotes axon branching whereas the latter promotes axon elongation. Each of these functions is blocked by the appropriate antibodies. This is the first demonstration that these two molecules, thought to exist in a complex in native matrices, can separately influence neurite outgrowth. Since the effects of heparan sulfate proteoglycan occur more slowly than laminin effects, they may have been missed by previous investigators. These results suggest that numerous molecules may influence neurite outgrowth and may yield different patterns of axonal branching. This has significant implications for neurite guidance by environmental cues since different matrix molecules may result in different pathways of axonal navigation. *Marianne Bronner-Fraser*

Structural Requirements for the Stimulation of Neurite Outgrowth by Two Variants of Laminin and Their Inhibition by Antibodies

D. Edgar, R. Timpl, and H. Thoenen
J. Cell Biol., 106, 1299—1306, 1988 5-19

Neurons extend neurites when cultured on substrates of specific cell types such as glia, or if the substrate contains molecules arising in certain cell types. The authors compared the laminins of Engelbreth-Holm-Swarm (EHS) tumor and RN22 Schwannoma cells, both of which stimulate rapid neurite outgrowth, suggesting a common neurite-promoting site.

Anti-laminin antisera inhibit only the activity of EHS laminin. The blocking antibodies are directed against the terminal heparin-binding domain of the laminin long arm. Immunoblot studies showed these epitopes to be absent in RN22 laminin. Antibodies against a 25-kDa proteolytic fragment of the long arm of laminin did not cross-react with native laminin, but they recognized the B chains of denatured EHS and RN22 molecules on immunoblots. The antibodies also bound to the long arm of laminin containing the neurite-promoting site, inhibiting its activity.

Normally nonantigenic B-chain sequences are present within or near the neurite-promoting site of laminin. The inability of anti-25-kDa fragment antibodies to recognize their epitopes in EHS or RN22 laminins is consistent with the view that the neurite-promoting site is not antigenically active in the native molecules.

♦Laminin is a large glycoprotein with numerous putative binding sites for cells and other extracellular matrix molecules. Since neurons respond readily to laminin substrates, there has been a great deal of interest in defining the region of neurite binding on the laminin molecule. Laminin consists of one A chain and two B chains, which form a cross-like molecule as visualized by rotary shadowing. The long arm of laminin in the region of the heparan binding site has been thought to represent the neurite binding site, since antibodies against this region block neurite outgrowth. The heparin binding region lies on the A chain.

This paper by Edgar et al. explores the region of laminin involved in neurite outgrowth by using several different types of laminin. The most significant finding is that laminin from a Schwannoma cell line which virtually lacks the A chain (and thus the heparin binding region) retains its capacity to support neurite outgrowth. Antibodies against the heparin binding region have no effect on promotion of neurite outgrowth by this Schwannoma-derived laminin. This suggests that the neuron binding site is on the B chain of laminin. The likely explanation for the blockage of neurite outgrowth observed with intact laminin by antibodies against the heparin binding site is by means of steric hindrance. These results help to elucidate the regions of laminin involved in neurite-promoting activity and explain many previous and apparently contradictory results about laminin and neurite outgrowth. *Marianne Bronner-Fraser*

Response of Sensory Neurites and Growth Cones to Patterned Substrata of Laminin and Fibronectin *in Vitro*
R. W. Gunderson
Dev. Biol., 121, 423—431, 1987 5-20

The glycoproteins laminin and fibronectin may be involved in the development and regeneration of the nervous system. Both enhance neurite elongation, and laminin may provide a preferred substrate for this process. The authors studied the responses of sensory neurites and growth cones to various patterns of laminin or fibronectin, applied to a background substrate of type IV collagen. Dorsal root ganglia from 5-d chick embryos were used in the studies.

Neurite elongation was restricted to a laminin pattern, not a fibronectin pattern. Neuritic preference for laminin was less evident when polylysine was included in the background substrate. Laminin also enhanced neurite defasciculation and stabilized growth cone protrusions.

Substrate-adsorbed laminin — but not fibronectin — provides areas of preferred substrate for neurite elongation when the background substrate is type IV collagen. Regional variations in laminin distribution suggest a possible role in establishing preferred paths of neurite elongation in the peripheral nervous system. Studies using interference reflection microscopy would help show whether adhesive as well as cytoskeletal effects are involved.

♦Sensory axons have been shown to respond to both fibronectin and laminin *in vitro* on planar substrates by extending neurites. In contrast to tissue culture substrates, matrices in the embryo exist in complex three-dimensional arrays which contain multiple extracellular matrix molecules which interact with one another. Therefore, it is necessary to study the nature of growth cone-matrix interactions on multiple matrix components in order to understand neurite outgrowth *in situ*.

In this study, Gunderson performs a careful analysis of growth cone interactions with different combinations of extracellular matrix substrates. He finds that sensory axons prefer and orient on laminin substrates when applied to a background of collagen IV. In contrast, axons show no preference for fibronectin over collagen IV. When polylysine is included with the collagen IV background, axons show no preference for laminin. Interestingly, direct adhesion measurements show that axons are less adhesive to laminin than other substrates. The significance of these findings lies in their attempt to compare and contrast many "adhesive" molecules which are possibly involved in neurite outgrowth. The results clearly demonstrate that axons have a hierarchy of substrate preferences, and suggest that a complex three-dimensional matrix may promote axonal guidance differently than a simple two-dimensional substrate. It is clear that laminin pathways can guide axons on the proper background matrix. The finding that axons are actually less adhesive on laminin than on other substrates highlights the possibility that cell-matrix interactions must be constantly made and broken during axon guidance. Thus, a less adhesive substrate may facilitate axon growth whereas a more adhesive substrate may actually prevent growth cone movement. *Marianne Bronner-Fraser*

The Role of Laminin and the Laminin/Fibronectin Receptor Complex in the Outgrowth of Retinal Ganglion Cell Axons
J. Cohen, J. F. Burne, C. McKinlay, and J. Winter
Dev. Biol., 122, 407—418, 1987

5-21

As the avian optic pathway develops, retinal ganglion cells (RGCs) send axons to the tectum along highly stereotyped routes restricted to the outer margins of the embryonic CNS. Immunohistochemical studies suggest that laminin, a component of the extracellular matrix, may act as a substrate for axonal growth in the developing vertebrate nervous system. Cultured chick RGCs extend neurites on laminin at an early stage of embryonic development.

Affinity-purified rabbit anti-laminin antibodies and monoclonal antibody JG22 were used, with a sensitive biotinylated antibody-streptavidin detection method, to determine whether laminin and JG22 antigen are expressed in the optic pathway of chick embryos. Laminin was found to be expressed on neuroepithelial cell processes before and during the early phases of RGC axonal outgrowth. As the response to laminin by RGCs was lost, laminin immunoreactivity became confined to basement membranes. When optic axons with neurofilaments were present in the pathway, the fibers expressed laminin/fibronectin receptors recognized by JG22 antibody. This staining persisted throughout embryonic development. JG22 antibody blocked neurite outgrowth on a laminin substrate.

It appears that laminin may define a transient preformed pathway that is specifically recognized by early RGC growth cones as they move toward their central target. Differences in neurite outgrowth responses over time may reflect the transition from a target-independent to a target-dependent phase of RGC development.

♦Laminin is thought to be important for axonal guidance since it can readily support neurite outgrowth from both peripheral and central neurons in tissue culture. Although laminin is prevalent in the peripheral nervous system, little laminin has been identified in the brain. Furthermore, some central nervous system-derived neurons appear to be sensitive to laminin for limited times during their development. In the avian retina, for example, embryonic day 6 axons migrate on laminin whereas embryonic day 11 axons do not. This suggests that neuron sensitivity to guidance by laminin may be temporally restricted.

This paper by Cohen et al., examines the distribution of laminin in the optic tract during the time of retinal axon ingrowth. The authors find a transient expression of laminin which correlates with the time that retinal ganglion cells send their processes through the optic pathway. Later laminin immunoreactivity becomes restricted. The retinal ganglion cells express an integrin receptor which binds to fibronectin and laminin. The authors find that antibodies against this receptor block the ability of E6 retinal ganglion cells to navigate on laminin substrates *in vitro,* suggesting that the neurons utilize this receptor to interact with laminin. These results represent one of the few examples where laminin has been found in the central nervous system. The transient expression of this matrix molecule during a window of time which correlates with axon ingrowth represents strong correlational evidence for the involvement of laminin in axon guidance *in vivo.* Furthermore, the authors demonstrate that retinal ganglion cells can interact with laminin by means of a specific cell surface receptor. These studies suggest that cell-matrix interactions may be important for the formation of proper neuronal connections. *Marianne Bronner-Fraser*

A Monoclonal Antibody against a Laminin-Heparan Sulfate Proteoglycan Complex Perturbs Cranial Neural Crest Migration In Vivo

M. Bronner-Fraser and T. Lallier

J. Cell Biol., 106, 1321—1329, 1988 5-22

Molecules of the extracellular matrix, such as fibronectin and laminin, promote neural crest migration *in vitro* and are present along neural crest migratory paths *in vivo.* Antibodies against the cell-binding region of fibronectin and against integrin, a cell surface receptor for fibronectin and laminin, interfere with normal cranial neural crest migration. Studies in avian embryos suggest that laminin and heparan sulfate proteoglycan (HSPG) exist as a complex. The possible role of this complex in neural crest migration was studied *in situ* using monoclonal antibody INO (inhibitor of neurite outgrowth) and chick embryos.

Injection of INO lateral to the mesencephalic neural tube led to a chiefly unilateral distribution, with binding in the basal laminae surround the neural tube, ectoderm, and endoderm, as well as in the cranial mesenchyme ipsilaterally. An identical staining pattern was observed with antibodies to laminin or HSPG. Injected antibody was detected for 18 h. Injection of embryos ranging from the neural fold stage to the 9-somite state resulted in ectopic neural crest cells outside the neural tube; crest cells within the tube lumen; and deformities of the neural tube. Injection of antibody against laminin or HSPG had no such effects. INO did not alter embryos having ten or more somites. Immunoprecipitation of INO antigen from 2-d chick embryos yielded a band characteristic of laminin.

These findings indicate that a laminin-PSPG complex has a role in neural crest migration *in vivo*. It is likely that many cell surface and cell substrate molecules are involved in guiding cell movement during neural crest migration.

♦Laminin is a large glycoprotein which is present in the basal laminae surrounding most epithelial tissues as well as in the interstitial matrices of some mesenchymal tissues. Recently, the importance of this molecule in cell and neurite guidance has been suggested by *in vitro* studies which demonstrate that laminin supports neurite outgrowth and regeneration. However, little is known about the function of laminin *in situ*. In it native state, laminin is thought to exist as a complex with heparan sulfate proteoglycans, as well as other molecules such as nidogen/entactin and collagen type IV.

This study by Bronner-Fraser and Lallier utilizes an antibody against a laminin-heparan sulfate proteoglycan complex which was isolated for its ability to inhibit interactions between neurons and laminin-containing complex substrates. The authors microinject this antibody into regions through which cranial neural crest cells migrate in order to interfere with cell binding to laminin-heparan sulfate proteoglycan within the extracellular matrix. They find that cranial neural crest cells undergo improper initiation of migration after antibody injection. Rather than following their normal pathways of migration, some neural crest cells come to lie external to the neural tube, whereas others are trapped within the lumen of the neural tube. These results suggest that neural crest cell interactions with laminin-heparan sulfate proteoglycan are necessary for normal cranial neural crest migration. This represents the first direct evidence demonstrating that a native laminin-containing complex is important for development *in vivo*. In addition, the authors have provided the first immunoprecipitation data characterizing the complex identified by the antibody. *Marianne Bronner-Fraser*

Neurofascin: A Novel Chick Cell-Surface Glycoprotein Involved in Neurite-Neurite Interactions

F. G. Rathjen, J. M. Wolff, S. Chang, F. Bonhoeffer, and J. A. Raper
Cell, 51, 841—849, 1987

5-23

Understanding of the molecular mechanisms of neurite-neurite interactions is very deficient. The authors describe a novel neurite-associated surface glycoprotein, termed neurofascin, that is expressed chiefly in neurite-rich areas of the developing chick nervous system. Antibodies against neurofascin inhibit the *in vitro* fasciculation of retinal neurites and the extension of sympathetic axons on axons.

Neurofascin is defined by reactivity with monoclonal antibody F6, which detects two polypeptides, 160 kDa and 185 kDa, in immunotransfers of brain plasma membrane proteins. Immunoaffinity chromatography using immobilized F6 yielded a group of related polypeptides. Neurofascin was expressed primarily in fiber-rich regions of cerebellum, spinal cord, and retina in the developing chick nervous system. Fab fragments of polyclonal antibodies to neurofascin interfered with the outgrowth of both retinal and sympathetic neurons *in vitro*. Studies with polyclonal antibodies to other antigens indicated that neurofascin is a unique glycoprotein in the chick nervous system.

Neurite-neurite interactions probably are mediated by many cell-surface glycoproteins, one of which is neurofascin. Current studies may indicate whether these molecules act independently or in concert with one another.

♦Neurons elicit long axonal processes which extend to their target sites. To date, little is known about what guides these neurites to specific targets. However, it is clear that neurites interact with one another, forming extensive axon bundles. This fasciculation of neurites may aid in proper neurite guidance, since neurites that are extended at later times could fasciculate with and follow other pioneering neurites to the proper target sites. Several cell surface adhesion molecules have been identified on neurons and/or their processes. However, the role of these molecules in neurite-neurite interactions remains poorly understood.

In this paper, Rathjen and colleagues characterize a novel molecule called neurofascin which is involved in neurite-neurite interactions. The distribution of neurofascin in fiber-rich regions of the brain and spinal cord is consistent with this molecule playing a role in neurite adhesion. Furthermore, antibodies against neurofascin inhibit bundling of retinal axons *in vitro* and decrease the length of sympathetic neurites grown on axonal substrates. The latter effect is substrate specific, since the antibodies have no effect on sympathetic neurites grown on laminin. These results demonstrate that: (1) neurofascin is a newly described molecule distinct from other molecules involved in axon-axon interactions; (2) neurofascin is distributed in axonal bundles; and (3) neurofascin is functionally important for neurite interactions. These findings, together with the previous identification of molecules present on neurites, show that numerous molecules are involved in neurite-neurite interactions, and suggest redundancy in this developmental mechanism. It is not yet clear whether these different molecules act independently or synergisticly. *Marianne Bronner-Fraser*

Occupation of the Extracellular Matrix Receptor, Integrin, Is a Control Point for Myogenic Differentiation

A. S. Menko and D. Boettiger

Cell, 51, 51—57, 1987 5-24

The integrins are a group of cell surface receptors that interact with components or the extracellular matrix. One example, defined by reactivity with the monoclonal antibody CSAT, is an integral membrane protein receptor complex in the chicken, consisting of alpha and beta polypeptide subunits. The surface expression of CSAT on myogenic cells changes as the cells differentiate in culture. The authors found that blocking integrin in myoblasts by binding of CSAT antibody reversibly blocks their ability to differentiate into myotubes.

Myogenic cultures were prepared from breast skeletal muscle of 11-d chick embryos. When replicating myoblasts were treated with CSAT, myoblasts continued to replicate and did not fuse or produce muscle-specific meromyosin. The block to normal differentiation was reversed by removing the antibody. Reversed cells had increased desmin and meromyosin, and they fused to produce contracting myotubes. CSAT antibody did not block the fusion of myogenic cells. It blocked myogenic differentiation at the myoblast stage.

The simplest explanation of these findings is that myogenic cells require a signal to initiate the terminal phases of myogenesis, a signal that is supplied by the interaction of integrin with the extracellular matrix. The matrix may provide an environment permitting specific differentiation to occur, rather than acting directly as an inducer.

♦During muscle differentiation, mononucleated myoblasts stop dividing, become aligned, and fuse to form syncytial myotubes. Following fusion, myotubes produce muscle specific proteins such as meromyosin. Thus, fusion is closely followed by changes in gene expression. The process of myoblast fusion involves cell-cell interactions and is also thought to be influenced by cell-matrix interactions.

Menko and Boettiger have examined the role of a cell surface fibronectin/laminin receptor in myoblast fusion. They find that antibodies against the "integrin" receptor block myoblast fusion. When fusion is blocked, so are the associated changes in gene expression. This alteration is completely and rapidly reversible by removal of the antibodies. These results show that blocking cell-matrix interactions can also block subsequent fusion and differentiation. This indicates that cell-matrix interactions are essential for myogenic differentiation, and represents one of the few direct examples illustrating that the extracellular matrix can influence phenotypic expression. Furthermore, the data suggest a direct link between extracellular signals and gene expression, mediated by the transmembrane integrin receptor. In this system, integrin represents a cell surface receptor which functions as a signal transducer between the extracellular matrix and the nucleus. *Marianne Bronner-Fraser*

N-Cadherin and Integrins: Two Receptor Systems That Mediate Neuronal Process Outgrowth on Astrocyte Surfaces
K. J. Tomaselli, K. M. Neugebauer, J. L. Bixby, J. Lilien, and L. F. Reichardt
Neuron, 1, 33—43, 1988 5-25

Receptor-mediated interactions between neurons and astroglia presumably are critical in the growth and guidance of CNS axons. The authors used antibodies to neuronal cell surface proteins to identify two receptor systems that mediate neurite outgrowth on cultured rat cortical astrocytes. Astrocytes express, *in vitro,* the neurite outgrowth-promoting factors to which both central and peripheral neurons can respond. It has been proposed that defective CNS regeneration reflects an inability of the glail cells to promote regrowth.

N-Cadherin is a calcium ion-dependent cell adhesion molecule that functions in the outgrowth of neurites on astrocytes by E8 and E14 chick ciliary ganglion neurons. Beta-1-class integrin ECM (extracellular matrix) receptor heterodimers function less obviously in E8, and not at all in E14 neurite outgrowth on astrocytes. *N*-Cadherin and integrin β_1 antibodies together virtually eliminated E8 ciliary ganglion neurite outgrowth on cultured astrocytes.

These neuronal receptors probably help regulate axonal growth on astroglia *in vivo.* N-Cadherin and integrins probably are only two of many neuronal receptors functioning in neurite outgrowth. Interaction of the receptors with proteins that regulate cytoskeletal function suggests one way in which growth cone morphology and motility could be regulated.

♦During the formation of neuronal connections, both cells and extracellular matrix molecules have been proposed to elicit neurite outgrowth. In the central nervous system, astroglia are thought to be involved in axon guidance, presumably by neuronal cell surface receptor-mediated interactions. One astroglial-derived molecule is laminin, which has been proposed to be important for neuronal guidance.

In this study, Tomaselli et al. examine the role of both cell-cell interactions and cell-matrix interactions in neurite outgrowth. They use antibodies against *N*-cadherin (which mediates a Ca^{++}-dependent cell-cell interaction), integrin (which mediates cell-laminin interactions), and N-CAM (which mediates a Ca^{++}-independent cell-cell interaction). They find that neurite outgrowth from young ciliary neurons utilizes both integrin and *N*-cadherin adhesion systems, though the latter is dominant. In contrast, older ciliary neurons only utilize *N*-cadherin-mediated adhesion. These results suggest that multiple receptor systems are involved in complicated processes such as neurite outgrowth. This is especially interesting in the case of astrocytes, which both have cell surface ligands for *N*-cadherin, probably by means of a homophilic interaction, and secrete laminin into a cell-associated extracellular matrix. Thus, a single cell type can elicit multiple cues yielding a redundancy of mechanisms involved in neuronal outgrowth. There also appears to be a time-related alteration in adhesion systems, with the *N*-cadherin system being the sole system utilized by E14 ciliary neurons. This emphasizes the importance of timing in the establishment of proper neuronal connections. *Marianne Bronner-Fraser*

Recognition of Position-Specific Properties of Tectal Cell Membranes by Retinal Axons *in vitro*
J. Walter, B. Kern-Veits, J. Huf, B. Stolze, and F. Bonhoeffer
Development, 101, 685—696, 1987 5-26

The authors developed an *in vitro* assay in which membrane fragments of two sources are arranged as a carpet of narrow alternating stripes. Axons growing on these carpets are confronted with the substrates simultaneously at the stripe borders, and will exhibit any preference for a given substrate through becoming oriented by the stripes and growing within the lanes of this substrate. The assay was used to study the growth of chick retinal axons on tectal membranes. Tissue strips from different areas of the retina were explanted, and the extending axons interfaced with stripes of cell membranes from areas of the optic tectum.

Nasal and temporal axons grew well on membranes of both posterior and anterior tectal cells. Temporal axons exhibited a distinct preference to grow on membranes of the anterior tectum. Nasal axons exhibited no preference. Positional differences in tectal membrane properties were evident only during development of the retinotectal projection. They were independent of tectal innervation by retinal axons. The anterior-posterior differences disappeared by embryonic day 14.

The graded property of chick tectal membranes may function in development of the topographic map. The gradient has the correct orientation, and the difference between anterior and posterior membranes is apparent only while the retinotectal projection is developing. This *in vitro* system may be used to study chemically modified membrane fragments.

♦In the formation of connections between the retina and tectum, a characteristic retino-tectal map is reproduced on the tectal surface which reflects the original positions of the retinal ganglion cells in the visual field. One explanation for the formation of this precise map is that positional cues may exist both on the retinal cells and on the surface of the tectum.

This study by Walter et al. devises a novel assay system to test such positional differences. Alternating narrow strips of membranes from two different tectal regions are placed on a substrate. Retinal axons from distinct regions of the retina are plated onto these substrate and the pathways followed by their axons are monitored. Many interesting findings emerge from this approach. First, the authors find that temporal axons have a marked preference for membranes from anterior portions of the tectum. They also find that an abrupt positional border distinguishes temporal from nasal axons. In the tectum, the differences between anterior and posterior tectum appear to be transient, corresponding to the time of axon ingrowth, and (in an accompanying paper) the authors show that posterior membranes inhibit temporal axon growth compared with anterior membranes. Thus, there appears to be an anterior to posterior gradient of inhibitory information. These are important findings which begin to define the phenomenological behavior of discrete groups of axons on regional tectal membranes. The extremely clever assay system establishes a basis for biochemical characterization of the relevant molecules responsible for axon preference. The results also have significance for theoretical modeling of the retino-tectal system, since they demonstrate the existence of a tectal gradient but suggest that no such gradient exists along the nasotemporal axis of the retina. Rather, there is an abrupt, nongraded change from nasal to temporal type axons. *Marianne Bronner-Fraser*

Cellular Expression of Liver and Neural Cell Adhesion Molecules After Transfection with Their cDNAs Results in Specific Cell-Cell Binding

G. M. Edelman, B. A. Murray, R.-M. Mege, B. A. Cunningham, and W. J. Gallin

Proc. Natl. Acad. Sci. U.S.A., 84, 8502—8506, 1987 5-27

Cell adhesion molecules (CAMs) of differing specificity have important roles in morphogenesis, histogenesis, and regeneration. Specific CAM binding functions may correlate with particular cell states, as indicated by recent studies of the structure of cDNAs specific with the Ca-dependent liver CAM (L-CAM) and Ca-independent neural CAM (N-CAM). The authors used vectors constructed with cDNAs specifying L-CAM and N-CAM chains to transfect liver cells ordinarily free of these molecules. The vectors contained the simian virus 40 early promoter and cDNA sequences encoding chicken L-CAM or the major polypeptide forms of chicken N-CAM.

Transfected cells expressing L-CAM showed uniform surface expression of the molecule. The cells aggregated readily, unlike nontransfected L cells. Aggregation was inhibited by Fab′ fragments of antibodies to L-CAM, but not by fragments of anti-N-CAM. Transfected cells spread relatively efficiently in culture. Transfected L cells expressing either of two polypeptide chains of N-CAM aggregated specifically with bound membrane vesicles from chick brain, and this was inhibited by Fab′ fragments of anti-N-CAM antibodies. Cells transfected with vectors specifying the third chain of N-CAM showed no phenotypic changes and no evidence of linkage of the chain to the cell membrane.

Different CAMs have specific roles in ligating the cells that synthesize them. The different phenotypic changes associated with specific CAMs suggest that CAM synthesis, or differing associations of CAM carboxyl-terminal domains with the cell surface and cortex, may lead to specific changes in the cells bound by that CAM. Transfection studies potentially can clarify many problems related to cell adhesion.

♦The neural cell adhesion molecule (N-CAM) and the liver cell adhesion molecule (E-CAM) have a wide distribution during embryogenesis and are thought to play a role in cell-cell adhesion during morphogenesis and histogenesis. Several strategies have been attempted to determine the function of these cell adhesion molecules in development. One approach is to perturb the function of the molecule using specific inhibitors. A complementary but less accessible approach is to express the molecule of interest in cells which normally lack the molecule and examine the affects on development.

This paper by Edelman et al., uses the latter approach to examine the function of N-CAM and L-CAM. Mouse cells which normally do not produce these CAMs are transfected with vectors containing cDNA sequences encoding for chick L-CAM and different forms of chick N-CAM. The transfected cells undergo changes in adhesion and morphology consistent with the presence of additional adhesion molecules on their surfaces. Cells expressing the sd of ld chains of N-CAM exhibited Ca^{++}-independent adhesion to each other and to brain vesicles. These interactions are specifically blocked by antibodies against N-CAM. Similarly, cells expressing L-CAM adhere to one another in a Ca^{++}-dependent manner which is specifically inhibited by antibodies against L-CAM. These results yield several important pieces of information. First, transfection of cells with specific cell adhesion molecules changes the morphological character of the cells as well as their interactions. Second, it is possible to create cell lines with heritable incorporation of the transfected cDNA. Third, the cells' biosynthetic machinery is able to synthesize and incorporate the protein products of the cDNA in a apparently normal and functional configuration. This approach opens the possibility of looking at the effects on cell behavior of adding functional cell surface molecules. *Marianne Bronner-Fraser*

Tolerance Induced by Thymic Epithelial Grafts in Birds
H. Ohki, C. Martin, C. Corbel, M. Coltey, and N. M. Le Douarin
Science, 237, 1032—1035, 1987 5-28

The thymus is central to immunologic discrimination between self and nonself. To examine the establishment of tissue tolerance during development, wing buds were grafted between day 4 embryonic histoincompatible chicks.

The allogenic wing was rejected 1 to 5 months after hatching. Therefore, these early embryonic grafts induced partial tolerance. If the allogenic wing bud was grafted from a quail embryo to a chick embryo, the wing was rejected within the first 2 weeks after hatching. However, grafts of quail thymic epithelium, prior to colonization by hemopoietic cells, induced tolerance of the quail wing graft. Tolerant birds were sacrificed at 1.5 to 3 months and histological examination of the thymus was carried out. Some thymic lobes in each tolerant chick were chimeric, with typical quail B-L immunoreactivity. Chick lymphocytes were differentiating in a quail thymic epithelial environment. In all lobes, the dendritic, macrophage B-L-positive cells were of host (chick) origin.

The maintenance of the foreign grafts in the presence of grafted thymic epithelium demonstrates the role of this cell type in the process of self/nonself discrimination.

♦Immunological recognition of self vs. nonself is a developmental process in which the thymus plays a central role. Previous investigations by LeDouarin and colleagues have shown that interspecific grafts of nervous system and some other tissues are tolerated during the embryonic period, but undergo rejection after development of the host immune system. Thus, rejection occurs even though the foreign tissue was incorporated into the embryo from very early developmental stages.

Here, Ohki et al. show that skin or limb bud grafts performed between young quail and chick embryos or between outbred strains of chickens are rejected postnatally. However, if part of the thymic rudiment from the donor is also transplanted to the host, then tolerance to the skin or limb graft is induced. This is an important finding which suggests that the presence of chimeric cells in the thymus is essential for recognition of the donor tissue as self. Thus, the thymus appears to be major tissue responsible for self vs. nonself immunological recognition. If this tolerance to tissue grafted during early embryogenesis extends to that adult, this type of thymic transplant could have important medical applications for organ transplantation research. *Marianne Bronner-Fraser*

The Effect of Thymus Environment on T Cell Development and Tolerance

P. Marrack, D. Lo, R. Brinster, R. Palmiter, L. Burkly, R. H. Flavell, and J. Kappler

Cell, 53, 627—634, 1988 5-29

T cell receptors react with foreign antigen only in the context of MHC products (e.g., IE of the mouse). During T cell maturation in the thymus, self-restriction occurs. In this process, thymocytes are selected if their receptors can recognize antigen in association with MHC products expressed in the thymus. However, T cells are eliminated if their receptors react with self-MHC. To resolve the paradox of positive and negative selection for reaction with self-MHC, it has been proposed that positive selection picks out thymocytes with affinity for self-MHC and that negative selection by bone marrow cells eliminates cells with high affinity for self-MHC. This would allow cells with a bias to recognize antigen in association with MHC products to circulate. This hypothesis was tested in mice.

Bone marrow from Vbeta17a+ F1, IEk-positive mice was grafted into B10.Q, H-2q, IE–, or IEk+ mice. Peripheral lymph node T cells were tested for expression of Vbeta8 as a control and for Vbeta17a. The percent of T cells expressing Vbeta17a was lowered to the same degree in either type of chimera. This demonstrated that IE expression on bone marrow cells was sufficient to cause deletion of Vbeta17a-bearing T cells. Induction of tolerance of MHC on bone marrow derived cells is therefore, due to clonal deletion. Animals expressing IE exclusively on thymus stromal cells were produced by irradiation and grafting or by transgenic methods. The expression of Vbeta17a was examined in these animals. The results from either type of mouse were the same. MHC expression by thymus epithelium was not sufficient to induce tolerance to that MHC. However, the grafted thymus epithelium survived in these animals. Expression of an MHC protein by thymus epithelium resulted in underselection for T cells that reacted with that molecule.

These results are not consonant with the standard explanation for selection of T cells. Therefore, an alternative model is proposed. In this mode, MHC molecules on thymic epithelium are different from those on bone marrow-derived cells. Thymocytes would be selected for self-restriction by reaction between their receptors and epithelial MHC peptides. These receptors would not interact with MHC molecules on bone marrow-derived cells. This theory could account for all the results presented in this paper.

♦The issue in question is the establishment of the T cell repertoire during thymic development of immunocompetent T cells and the apparent paradox of self-MHC restriction in linked recognition of antigen associated with "self" major histocompatibility complex (MHC) products vs. tolerance of self-MHC. It has generally been accepted that thymocytes are both positively and negatively selected for recognition of self-MHC, with positive selection of cells, which will recognize antigen in the context of self-MHC, through both high- and low-affinity interaction of thymocytes with MHC products expressed on cortical epithelial cells of the thymus and subsequent negative selection of cells marked for deletion through high-affinity thymocyte binding to MHC products expressed on bone marrow-derived cells in the thymic medulla. Selection thus produces T cells which

are weakly reactive toward self-MHC alone but which can be strongly reactive toward antigen plus self-MHC. Both selection processes occur in the absence of overt antigen. The hypothesis has been (re)tested by Marrack and colleagues by tracking Vβ17a (in the T cell antigen receptor's β chain) with a monoclonal antibody and taking advantage of its correlation with the IE-reactivity of T cells that bear it. It was initially established that the expression of IE on bone marrow-derived cells in reconstituted, irradiated mice was sufficient to induce clonal deletion of $V\beta17a^+$ T lymphocytes, supporting the negative selection portion of the hypothesis. However, using mice expressing IE on thymus stromal cells but not elsewhere (i.e., bone marrow-derived cells) which were constructed by thymus lobe grafts or by transgeneic approaches, it was demonstrated that the expression of MHC gene products on thymic epithelium failed to induce the production of the large numbers of IE-reactive T cells predicted on the basis that positive but not negative selection was allowed. The IE-reactive T cells were in fact produced in lower numbers than those in mice which were totally negative for IE. Moreover, their response in reactions with IE-positive cells was not as strong as the response of control T cells of the same MHC type which had developed in normal syngeneic thymuses. The investigators propose a unifying hypothesis to account for their results and those of others. The hypothesis is based on inherent differences between MHC molecules on thymic epithelium and on bone marrow-derived cells, a difference created by the association of the MHC molecules with specific self-derived peptides. Thus, the "altered" MHC products would be read differently by the maturing T cell. For example, recognition of $(\text{peptide})_a$MHC on the stromal cells would induce self-restricted antigen-responsive cells which might escape deletion because they could not see $(\text{peptide})_b$MHC on bone marrow-derived cells. The postulate dissociates the processes of positive and negative selection, and it is supported by the discovery of an unidentified peptide in the putative antigen-binding site of a crystalline class I MHC molecule (Bjorkman et al., *Nature (London)*, 329, 506—512, 1987).
Judith A. K. Harmony

The Adult T-Cell Receptor δ-Chain is Diverse and Distinct from That of Fetal Thymocytes

J. F. Elliott, E. P. Rock, P. A. Patten, M. M. Davis, and Y.-H. Chien
Nature (London), 331, 627—631, 1988 5-30

The T-cell receptor (TCR), carried on the surface of T-cells in association with the CD3 polypeptide complex, recognizes foreign molecules. This property is largely contained in the TCR alpha-beta heterodimers. Another CD3-associated heterodimer, gamma-delta, has been identified, but its function remains unknown. Developmentally, gamma-delta cells appear before alpha-beta cells and 1 to 10% of mature peripheral T-cells carry gamma-delta. In order to characterize the gamma-delta region and its expression, 17 delta-containing cDNA clones from adult double negative (CD4–8–) thymocytes were sequenced.

Only a limited number of the V-delta sequences were used in this adult population. There was little overlap between these sequences and those of V-alpha or fetal V-delta. However, the potential number of different delta chains that can be made is apporximately 10^{13}.

This calculation implies that there are approximately 10^{17} possible gamma-delta heterodimers. This may have important implications for T-cell recognition.

♦The 1980s brought an end to the search for the elusive T cell receptor for antigen (TCR), and the complete primary sequences of the subunits were deduced from cDNA and genomic clones. A surprising result was the discovery of two types of heterodimeric TCRs, $\alpha\beta$ and $\gamma\delta$, both associated with the accessory polypeptides designated CD3. The structures of the subunits are similar. The TCR-$\gamma\delta$ exists on a small fraction of peripheral blood T cells in the adult but is expressed on the majority of skin dendritic cells. The TCR-$\alpha\beta$ functions in the reactions of mature helper and cytotoxic cells. The function of TCR-$\gamma\delta$ is unknown and the subject of considerable debate, although it has been shown *in vitro* that TCR-$\gamma\delta^+$ cells can mediate the same effector functions as TCR-$\alpha\beta^+$ lymphocytes. During the development of murine fetal thymocytes, expression of the $\gamma\delta$ heterodimer precedes that of $\alpha\beta$ (Pardoll et al., *Nature (London),* 326, 79—81, 1987). Recently, Elliott and colleagues added to the TCR-$\gamma\delta$ mystery by finding that the V region sequences of δ differ significantly between fetal and adult murine thymocytes. The finding that fetal and adult thymuses contain different types of T cells suggests that a crucial aspect of the thymic environment in which the T cells develop is different in the fetal vs. adult thymus. The difference, which cannot be attributed to the MHC type, has obvious implications for thymic selection and should be the subject of vigorous scientific inquiry. *Judith A. K. Harmony*

Differential Splicing and Alternative Polyadenylation Generates Distinct NCAM Transcripts and Proteins in the Mouse

J. A. Barbas, J.-C. Chaix, J. Steinmetz, and C. Goridis

EMBO J., 7, 625—632, 1988 5-31

NCAM is a class of membrane glycoproteins that are important in development. The three prominent NCAM proteins migrate in SDS gels with apparent molecular weights of 180, 140, and 120 kDa. The expression of these peptides varies with cell type and age. Evidence indicates that these NCAM isotypes are all derived from one gene. The NCAM gene has been cloned, and the sequence of six exons, introns, and flanking sequence of the

3′ region of this gene is reported.

The mouse NCAM gene was isolated from overlapping cosmid clones. Southern blots using cDNA probes and Northern blots were used to identify exons. The 5′ region of the gene hybridized to all four prominent NCAM mRNAs. The 3′ region contained the part of the NCAM gene where alternative processing of the primary transcript to generate the different classes of mRNAs could occur. Therefore, approximately 9.5 kb of this region was sequenced. The second exon contained polyA sites for the two smaller transcripts. These two transcripts varied in the length of their 3′ noncoding regions and both encode NCAM-120. The second exon was excluded from the 7.4-kb transcript which encodes NCAM-180. The second and the fifth exon were not incuded in the 6.7-kb RNA, which encodes NCAM-140. Inverted repeats were found to surround exon 5, which could potentially form a stem-loop structure.

The 3′ portion of the mouse NCAM gene contains five exons. Differential splicing and alternative use of two polyA sites can account for the four prominent NCAM mRNAs.

♦The neural cell adhesion molecule (NCAM) is a membrane-associated glycoprotein that serves as a ligand in the formation of homophilic cell-cell bonds. In addition, it is thought that the molecule is involved in several aspects of early morphogenesis of the vertebrate embryo. For example, transient expression occurs in a variety of early embryonic structures such as nonmigrating neural crest cells, notochord, placodes (lens, otic, and pharyngeal), and mesonephric primordium. Continuous expression is found in the primitive neuroepithelium, differentiated central and peripheral nerve cells, developing skeletal muscle (which includes somites, dermomyotome, myoblasts, and myotypes), and developing cardiac muscle. Later in development, the molecule is found predominantly on neurons, glia, and muscle cells, and it is thought to participate in several types of interactions including bundling of nerves, guidance of axonal growth cones along glial pathways, and formation of neuromuscular synapses. NCAM is known to be a single polypepetide with structural heterogeneities in both the protein and carbohydrate portions of the molecule. For example, the amount of sialic acid by weight varies with respect to location and age of the embryo, and this observed heterogeneity is thought to represent different requirements for adhesion in the embryo. The polypeptide itself exists primarily in three molecular weight forms, the expression of which is also dependent on tissue source and age. *In vitro* translation studies have shown that the three polypeptide chains are synthesized from four to five different mRNAs, each of which exhibits developmental-stage and cell-type specificity in their expression. The mouse NCAM gene has been cloned by the Goridis group and has been shown to map to mouse chromosome 9. This present paper from the same group reports on the exon-intron structure of 3′ part of the mouse gene. This region contains six exons of which the most 5′ exon is expressed constitutively in all four classes of NCAM mRNAs. The second exon contains the poly (A) addition sites for the two small messages, both of which code for the smaller NCAM protein (120,000 kDa). This second exon is absent from the largest message which encodes the largest of the three NCAM proteins (180,000 kDa), and the second and fifth exons have been spliced out of the message that codes for the middle size NCAM protein (140,000 kDa). Thus, through differential splicing and alternative polyadenylation, three distinct NCAM proteins are generated from one gene. The question remaining to be addressed is what regulates the splicing and expression of the different forms of NCAM at different times during development. *Terry Magnuson*

A Novel Form of TNF/Cachectin Is a Cell Surface Cytotoxic Transmembrane Protein: Ramifications for the Complex Physiology of TNF

M. Kriegler, C. Perez, K. DeFay, I. Albert, and S. D. Lu
Cell, 53, 45—53, 1988 5-32

Tumor necrosis factor (TNF) is a 17-kDa peptide that has been implicated in tumor regression, septic shock, and cachexia. To explore the mechanism of induction of these three different states by one molecule, TNF from human monocytes was characterized.

Human monocytes were induced to synthesize TNF by treatment with lipopolysaccharide and the tumor promoter, PMA. A polyclonal antibody was used to identify TNF by Western analysis. Instead of detecting a 17-kDa antigen, these antibodies detected a 26-kDa protein. This protein was the size of the TNF protein, if the putative signal sequence was not removed. TNF cDNA was subcloned into a retroviral vector and transfected into NIH 3T3 cells to produce a constitutive TNF-producing cell line. Both the 17- and 26-kDa forms of TNF were produced by this cell line. Only the 17-kDa protein could be precipitated from cell supernatants, while the 26-kDa fraction was isolated from purified membranes. A pulse-chase experiment demonstrated that the 26-kDa protein was the precursor of the 17-kDa TNF molecule. *In vitro* translation yielded a pure preparation of 26-kDa TNF. This product was not cleaved in the presence of dog pancreatic microsomes. Alkali treatment, which opens microsomes without disrupting their membranes, did not release 26-kDa TNF, indicating that it was tightly associated with membranes. After protease treatment, the portion of the protein exterior to the membrane was digested, leaving a 24-kDa peptide. As all methionines are at the amino terminus of the protein, and ^{35}S-methionine-labeled protein could not be detected after protease digestion in these experiments, the amino terminus of the protein must protrude from the membrane. TNF inserted into membranes was cytotoxic to L929 cells.

These studies have demonstrated that TNF exists in two forms, a 26-kDa form and a 17-kDa form. There is a precursor/product relationship between these two forms. The 26-kDa form of TNF appears to be an integral membrane protein that must be cleaved to the 17-kDa form to be released into the serum. The conversion between these two forms may control the effects of TNF.

♦Monokines, secreted by monocyte/macrophages particularly when activated, such as interleukin 1 (IL1) and tumor necrosis factor (TNF) can have both localized and systemic effects of a diverse nature. For example, TNF is cytotoxic and can contribute to tumor regression and limit viral action and it can induce the cachexia associated with chronic illness as well as septic shock. Unraveling the molecular mechanisms by which reasonably simple peptides can have multiple biological activities is a challenge to molecular and cellular biologists. Kriegler and cohorts pose the interesting possibility that the distinct functions of TNF may depend on whether TNF exists as a 26-kDa transmembrane protein or as a 17-kDa extracellular protein derived from the 26-kDa precursor. In monocytes activated by lipopolysaccharide (LPS) and a tumor promoter (PMA), immunoprecipitated TNF was the 26-kDa form in which the putative 76-amino-acid signal sequence was not removed. NIH 3T3 cells transfected with TNF cDNA produced both 26-kDa and 17-kDa TNF; in pulse-chase experiments, cell 26-kDa protein was converted to secreted 17-kDa protein. Both 26-kDa and 17-kDa TNF synthesized *in vitro* from cDNAs were cytotoxic to L929 cells; however, the 26-kDa protein had to be properly oriented in microsomal membranes to be biologically active. The required orientation exposed the C terminus. Membrane and secreted forms of epidermal growth factor, transforming growth factor-alpha and interleukin 1 have been reported, suggesting that multiple sites of residence may be a common theme in the physiology of bioactive peptides. Kriegler et al. hypothesize that the membrane-intercalated and secreted forms of TNF function in fundamentally different ways, an hypothesis which must now be tested. Also, the mechanism of "secretion" of the 17-kDa form of TNF must be resolved. *Judith A. K. Harmony*

Lymphocyte Function-associated Antigen-1 (LFA-1) Interaction with Intercellular Adhesion Molecule-1 (ICAM-1) is One of At Least Three Mechanisms for Lymphocyte Adhesion to Cultured Endothelial Cells

M. L. Dustin and T. A. Springer
J. Cell Biol., 107, 321—331, 1988 5-33

Monoclonal antibodies to LFA-1 inhibit lymphocyte adhesion. ICAM-1 is a putative LFA-1 ligand. To test the function of ICAM-1 in lymphocyte adhesion, resting and activated lymphocytes were exposed to ICAM-1 in reconstituted planar membranes.

ICAM-1 expression was assayed on endothelial cells by radioimmunoassay and immunofluorescence flow cytometry. Basal expression of ICAM-1 varied between 5 to 10 × 10^4 sites per cell. After stimulation, there were approximately 3.5 × 10^6 sites per endothelial cell. Both the JY B lymphoblastoid cell line and T lymphoblasts adhered to ICAM-1 in reconstituted planar membranes. Adhesion was inhibited by antibody to ICAM-1 or LFA-1. Adhesion of JY cells, normal and LFA-1 deficient T cells, and resting peripheral blood lymphocytes to endothelial cell monolayers was also examined. Both LFA-1-dependent and LFA-1-independent adhesion was observed. LFA-1 adhesion was estimated to account for the majority of the cell-cell adhesion. The LFA-1 dependent cell adhesion could be subdivided into an ICAM-1-dependent component, which was increased by cytokines, and an ICAM-1-independent component, which was not affected by cytokines (see Figure 7, *J. Cell Biol.*, 107, 329, 1988).

ICAM-1 appears to be a ligand for lymphocyte cell adhesion. However, at least three lymphocyte adhesion pathways appear to exist. Only one of these pathways involves ICAM-1.

♦The attachment of a cell to other cells or to the extracellular matrix is a complex process, incompletely understood at the molecular level, but critically important in embryonic development and in the adult organism, e.g., organ function and tumor cell metastasis. In the lymphoid system, the adhesion of lymphocytes to endothelial cells precedes entry of circulating immature and immunocompetent lymphocytes into the lymphoid tissues and is essential to lymphocyte extravasation in immune and inflammatory responses. An understanding of the adhesion process and its regulation will offer the opportunity to comprehend the dynamic nature of lymphoid tissue and to manipulate the movement of its cells so as to control the body's immune response. The recent report by Dustin and Springer documents that the interaction between endothelial cell membrane ICAM-1 and lymphocyte membrane LFA-1 is important in adhesion but does not represent the only mechanism of adhesion between these two cell types. LFA-1, a heterodimeric member of the integrin superfamily of adhesion molecules, is expressed only on leukocytes whereas ICAM-1, a single-chain glycoprotein, is expressed on cells of numerous lineages. ICAM-1 is a member of the immunoglobulin supergene family, having five immunoglobulin-like domains and high homology with neural cell adhesion molecule NCAM (Staunton et al., *Cell*, 53, 927—933, 1988). Lymphocyte adhesion to endothelial cells and to purified ICAM-1/phospholipid recombinants was compared, testing unstimulated or activated T and B lymphocytes and unstimulated or activated cultured endothelial cells. Three classes of adhesive interactions were defined: (1) a basal LFA-1-dependent, ICAM-1-independent binding, (2) a LFA-1-, ICAM-1-dependent binding which was substantially upregulated by monokines and lipopolysaccharide (LPS), and (3) a LFA-1-independent pathway which was upregulated by monokines and LPS. The exact contribution of each pathway to the total adhesive interaction could not be assessed accurately by the assay utilized. It was estimated that LFA-1-dependent adhesion accounted for 60 to 90% of the total adhesion and the LFA-1/ICAM-1-dependent subset as much as 60% of this, depending on activation. Now that the pathways of interaction have been characterized descriptively, the intriguing and unanswered question is: What is the specific role of each of the adhesion interactions in lymphocyte trafficking and lymphocyte function? *Judith A. K. Harmony*

Structure of the Human Class I Histocompatibility Antigen, HLA-A2

P.J. Bjorkman, M. A. Saper, B. Samraoui, W. S. Bennet, J. L. Strominger, and D. C. Wiley

Nature (London), 329, 506—512, 1987 5-34

Human leukocyte antigens (HLA or class I histocompatibility antigens) are polymorphic membrane glycoproteins encoded within the major histocompatibility complex (MHC) and found on the surface of nearly all cells. HLA molecules are the targets of transplant rejection, and T cell receptors only recognize antigen in the context of HLA molecules. To gain an understanding of HLA structure and function, HLA-A2 was purified and crystallized with 3.5-Å resolution.

Soluble HLA-A2 was purified from papain digested plasma membranes of the cell line JY. The papain cleavage generates a molecule composed of alpha1, alpha2, alpha3, and beta2-microglobulin. This molecule was crystallized and a monoclinic electron density map was calculated to 3.5-Å resolution. HLA-A2 consists of two pairs of structurally similar domains. Alpha1 and alpha2 have the same tertiary fold, and alpha3 and B2M have the same tertiary fold (see Figure 2, *Nature*, 320, 508, 1987). The alpha3 and B2M domains are composed of two antiparallel beta-pleated sheets, connected by a disulfide bond. The alpha1 and 2 domains consist of antiparallel beta-pleated sheets, which are spanned by a long alpha-helical region. The contact between the alpha3 domain and B2M is different from that seen in antibody molecules. A groove that is 25 Å long and 10 Å wide lies between the alpha helices of the alpha1 and 2 domains (see Figure 6, *Nature*, 329, 509, 1987). This groove is on the surface of the molecule and is a good candidate for the antigen binding site. In this site, a region of electron density that could not be accounted for was observed. This may be due to antigen bound in the groove.

The HLA-A2 structure described above may be a good model for the interpretation of the structure of class I and class II histocompatibility antigens.

The Foreign Antigen Binding Site and T Cell Recognition Regions of Class I Histocompatibility Antigens

P. J. Bjorkman, M. A. Saper, B. Samraoui, W. S. Bennett, J. L. Strominger, and D. C. Wiley

Nature (London), 329, 512—518, 1987 5-35

The three-dimensional structure of HLA-A2 was determined, and a prominent groove was identified as the probable foreign antigen recognition site in the previous article. This groove was further characterized.

All the accessible residues in the vicinity of the binding site are listed in Figure 1 (see *Nature*, 329, 513, 1987). Their potential for making contacts is also indicated. The specificity of T cell recognition is determined by HLA polymorphism. Many of the polymorphic residues responsible for recognition by T cells and binding of antigen are located in this site (see Figure 3, *Nature*, 329, 515, 1987). A majority of amino acid substitutions that alter T cell recognition face into this site.

The discovery that molecules remained bound to the putative antigen recognition groove throughout purification and crystallization implies a strong interaction between HLA-A2 and this peptide. This suggests that HLA is always bound to a peptide and that displacement of this peptide may be necessary for cellular immunity.

♦Class I histocompatibility antigens (human leukocyte antigens, HLA), polymorphic glycoproteins encoded within the major histocompatibility complex and expressed on the surfaces of nearly all cells, are key players in the induction and development of immune responses. HLA molecules on antigen-presenting cells present antigen to a subset of T lymphocytes, and they are the targets of antibodies and cytotoxic T cells during foreign transplant rejection and lysis of virus-infected cells. Processed antigen bound to MHC is recognized by the T cell antigen receptor (TCR). How a limited number of HLA molecules expressed in an individual can interact with and present multitudes of foreign antigens has puzzled immunologists for decades. Class I MHC molecules are heterodimers, consisting of a transmembrane glycoprotein complexed to β2-microglobulin which is itself not membrane associated. Bjorkman and colleagues succeeded in crystallizing β2-microglobulin complexed with a fragment of HLA-A2, separated from the cell membrane by papain digestion, in which the α1, α2, and α3 domains of HLA-A2 were intact. Their analysis of the crystal structure (3.5 Å resolution) sheds considerable light on the antigen presentation problem. In the crystal structure, the paired membrane-proximal α3 HLA domain and β2-microglobulin resemble immunoglobulin constant domains but they interact in a manner distinct from that between pairs of constant domains in known antibodies. Distal domains α1 and α2 are nearly identical and are unlike constant domains, forming a platform structure of eight antiparallel β-pleated sheets topped by α-helices which together define a long (~25 × 10 Å) groove. The groove is proposed to be the antigen-binding site, and in the crystal it is occupied by an unidentified antigen. Most of the polymorphic residues and residues determined to be important in the interaction of MHC with antigen and in that between the MHC molecule and the TCR are situated about the groove. The groove theoretically can accommodate a number of different processed antigens which can form α-helices by recognizing some common antigen structural motif such as the peptide backbone. The TCR thus views a ligand comprised of one side of the helix of the antigen framed by the polymorphic residues of the MHC molecule. Polymorphic MHC residues, together with processed antigen, generate a sufficiently distinct surface to allow for specificity when presented to the TCR. *Judith A. K. Harmony*

Endoproteolytic Cleavage of gp160 Is Required for the Activation of Human Immunodeficiency Virus

J. M. McCune, L. B. Rabin, M. B. Feinberg, M. Lieberman, J. C. Kosek, G. R. Reyes, and I. L. Weissman

Cell, 53, 55—67, 1988 5-36

The envelope protein of HIV is synthesized as gp160 and then cleaved intracellularly, in a manner similar to other retroviruses, to gp120-gp41. Infection is initiated by binding of the envelope protein to a cell's CD4 receptor. It is possible that the cleavage step is necessary for envelope protein activity. Therefore, the ability of this protein to be cleaved was experimentally manipulated.

The predicted amino acid sequences of all characterized retroviral envelope proteins were compared. A conserved tryptic-like cleavage site (Arg-X-Lys/Arg-Arg) was observed between the larger and smaller subunits within these polyproteins. Within HIV gp160, the sequence Arg-Glu-Lys-Arg was found. Site-directed mutagenesis of this region was performed. The 30-mer mutating oligonucleotide (see Figure 5-36) introduced five novel base pairs, abolished the tryptic site, and introduced a chymotryptic site. The resultant virus was named RIP7/Mut10. RIP7/Mut10 synthesized proviral DNA and viral proteins and assembled virions; however, these virions were biologically inactive in a transfection assay with COS-1 cells. If these virions were treated with 1 to 4 μg/ml chymotrypsin, which allowed cleavage at the mutated site, biological activity was rescued.

Therefore, cleavage of HIV gp160 is necessary for activation of the envelope protein and viral infectivity. Interference with the intracellular proteolytic cleavage step is a possible method of combating infection with HIV.

♦Maturation or "activation" of the human immunodeficiency virus HIV is crucial to its pathogenicity. Delineation of the steps in the maturation

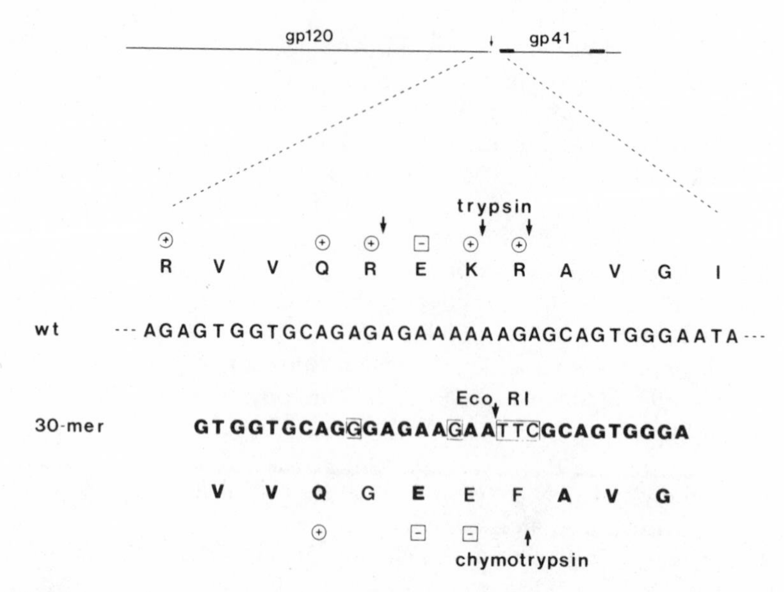

FIGURE 5-36. The mutating oligonucleotide used to remove the endoproteolytic cleavage site of gp160. A schematic of the HIV envelope protein is shown at the top, with the hydrophobic regions of gp41 depicted as darkened bars. The next two lines show the wild-type (wt) amino acid and nucleotide sequences in the vicinity of tryptic cleavage site #1 (see Figure 1, *Cell*, 53, 57, 1988). The mutating 30-mer is shown next, with the 5-bp change boxed and the new EcoRI site indicated. The three-codon change is depicted as a mutated amino acid sequence in the last line, with a novel chymotryptic cleavage site after phenylalanine. The one-letter amino acid code is used; positively (+) and negatively (–) charged residues are so designated. (From *Cell*, 53, 57, 1988. Copyright by Cell Press. With permission.)

process will allow the design of specific therapeutic inhibitors which are effective in preventing HIV infectivity and AIDS. Following the classic work of Choppin and collaborators on the infectivity of enveloped viruses, notably Sendai and influenza, which occurs via virus-cell fusion, McCune et al. have identified a single tryptic-like proteolytic cleavage site in the HIV envelope polyprotein gp160 which must be hydrolyzed, producing a heterodimer of gp120 and gp41, for the virus to be infective. The site, following a arg-X-lys/arg-arg sequence, is conserved among retroviruses. In the case of HIV, cleavage releases gp41 which contains a long stretch of hydrophobic residues at its N terminus that correspond to the fusion-inducing domains of the sendai F and influenza HA glycoproteins. The cleavage site was removed by site-directed mutagenesis and was replaced by a chymotrypsin-sensitive site in a vector carrying a biologically active HIV proviral clone and including the SV40 origin for replication and the xgpt gene for dominant selection. The native construct RIP7 and the mutated construct RIP7/mut10 were separately transfected into COS-1 cells. The results in both cases were almost indistinguishable based on the criteria of proviral replication, RNA processing, protein expression and viral assembly; however, virions produced from RIP7/mut10 were not infectious. The major difference between RIP7 ad RIP7/mut10 virions was the absence of gp120 and gp41 in the RIP7/mut10 virions. Generation of these products from the gp160 polyprotein by limited chymotrypsin digestion restored infectivity. It will now be important to identify the protease which normally cleaves gp160 and to determine whether it can be selectively if not specifically inactivated. *Judith A. K. Harmony*

Envelope Glycoprotein of HIV Induces Interference and Cytolysis Resistance in CD4⁺ Cells: Mechanism for Persistence in AIDS
M. Stevenson, C. Meier, A. M. Mann, N. Chapman, and A. Wasiak
Cell, 53, 483—496, 1988 5-37

HIV appears to play a direct role in the depletion of CD4+ helper T lymphocytes seen in AIDS patients. Persistently infected T lymphocyte cell lines have reduced expression of the CD4 receptor, which appears to be mediated by the virus. The ability of the HIV envelope glycoprotein to modulate CD4 expression was investigated.

Retroviral vectors, containing tat and art/trs, in addition to the HIV envelope protein gene under the control of the cytomegalovirus immediate early promoter, were used to insert and express the envelope protein gene in the CD4+ T cell line CEM. Integration of this provirus was accompanied by constitutive transcription of the HIV sequences. Fluorescence-activated cell sorter analysis indicated that CD4 expression on the cell surface was reduced by 60% in the infected cells. However, CD4 transcription and translation was not affected. Immunoprecipitation demonstrated an interaction between the CD4 protein and the viral envelope protein in these cells. The reduced CD4 expression on the surface of infected cells appeared to be due to the intracellular formation of a complex between the HIV envelope protein and the CD4 receptor. These cells were also resistant to syncytium formation and cytolysis following superinfection. After treatment with TPA (1 ng/ml/3 to 5×10^5 cells), which stimulates viral replication, the majority of these cells had lysed within 96 h. However, there was no evidence of syncytium formation.

Expression of the HIV envelope protein in CD4+ cell lines confers a cytolysis-resistant phenotype. A persistent viral infection, rather than cytolysis, is the result of HIV superinfection. Envelope protein-mediated interference coupled with superinfection may allow viral persistence and latency to occur during the course of AIDS infection. Factors that induce viral replication would interrupt latency and induce cytolysis, by a pathway that does not involve syncytium formation, and lead to CD4+ cell depletion.

♦Newly synthesized viral envelope glycoproteins of retroviruses and other enveloped viruses, e.g., adenovirus, can mask the expression of host membrane receptors. This phenomenon is termed interference. Human immunodeficiency virus (HIV), the etiologic agent of acquired immunodeficiency syndrome (AIDS), is linked to the depletion of $CD4^+$ T-helper lymphocytes, leaving the patient severely compromised immunologically. The basis of this depletion is unknown, and may involve multiple mechanisms. *In vitro,* HIV masks the expression of its receptor, the CD4 molecule, when it infects certain lines of $CD4^+$ T lymphocytes. Cells persistently infected with HIV are resistant to the cytopathic effects of the virus, suggesting that interference with CD4 expression may, at least in part, explain the phenomena of viral persistence and latency and immune dysfunction in the course of AIDS. Stevenson and associates have established that the expression of HIV envelope glycoprotein (gp160) in cytolysis-sensitive $CD4^+$ CEM T cells confers a cytolysis-resistant phenotype on the cells. A construct of the HIV DNA, including the entire envelope gene and the tat and art/trs genes required for efficient translation of the envelope glycoprotein, driven by the cytomegalovirus intermediate early gene promoter, was inserted into an amphotropic retrovirus vector PA317 and transfected into CEM cells; infected cells were selected in G418. In controls, the HIV tat gene only was included in the vector. Expression of gp160 and its processed product gp120 resulted in a 60% reduction in membrane CD4 without significant reduction in CD4 mRNA. Reduction in membrane CD4 was correlated with the intracellular formation of complexes between CD4 and gp160/120, indicating that interference of CD4 post-transcriptional processing and/or routing represented the primary mechanism of CD4 modulation in these infected cell lines. Control CEM cells and CD4-deficient CEM cells expressing the envelope glycoprotein were infected with HIV. Receptor reduction in gp160/120-positive CEM cells correlated with resistance to virus reinfection, as assessed by decreased accumulation of unintegrated virus DNA, and with greatly diminished virus cytopathicity. The reinfected gp160/120-positive cells were characterized by a sustained release of virus particles; reinfected control CEM cells were lysed. The authors thus propose that the envelope glycoprotein and its ability to modulate CD4 expression is an important factor in determining whether a lytic or persistent HIV infection is established. Viruses such as HIV which can diminish the expression of surface proteins necessary for normal immune responses may have a selective evolutionary advantage. *Judith A. K. Harmony*

Cell Lineage and Developmental Fate 6

INTRODUCTION

Throughout development, the ultimate phenotypic fate of any individual cell seems to be well regulated. The relationship between different cells as they differentiate into the same or different cell phenotypes is particularly interesting. The basis for cell determination, or the decision about which phenotype a cell will ultimately express, has intrigued developmental biologists for years.

Cell lineage relationships are influenced by several different factors. Specific genes encode proteins that can direct the lineage progression of a particular cell. Cell position relative to other cells and to the extracellular matrix can influence the ultimate fate of a cell, as can active induction (or repression) of a cell phenotype by neighboring cells.

Cell fate is somewhat "plastic" and can be modified to some degree. Finally, cell death can also provide important lineage signals.

The articles selected in this chapter all deal with either the genetic, the molecular, or the cellular aspect of cell lineage and developmental fate. This area of interest has been extensively investigated in the nematode C. *elegans,* where the circuitry of cell lineage relationships is known in elaborate detail.

A Lineage-Specific Gene Encoding a Major Matrix Protein of the Sea Urchin Embryo Spicule, I. Authentication of the Cloned Gene and Its Developmental Expression

S. Benson, H. Sucov, L. Stephens, E. Davidson, and F. Wilt

Dev. Biol., 120, 499—506, 1987 6-1

The sea urchin embryo forms calcareous endoskeletal spicules in gastrulation, and these increase in size and complexity as the pluteus larva matures. The proteins constituting the organic matrix of the spicule have been characterized. A polyclonal antibody that reacts specifically with matrix proteins has made it possible to screen an expression vector cDNA library in an attempt to identify cDNAs and genomic sequences that encode the spicule matrix proteins.

The authors have isolated a cDNA that encodes a predominant spicule matrix glycoprotein termed SM50. The cognate RNA transcript encodes a 50-kDa protein that is precipitated by polyclonal antisera against spicule matrix proteins, and is present only in polyadenylated RNA at stages making a spicule. The 2.2-kb mRNA was first detected at late cleavage stages and rapidly accumulated as the primary mesenchyme formed, peaking in the late gastrula and pluteus stages. Hybridization to cellular transcripts was first detected in primary mesenchyme cells as they entered the blastocoel. Transcripts remained confined to these cells during later development.

The spicule matrix protein gene provides a lineage-specific marker with which to study sea urchin development. It may be especially useful in studying determination and differentiation during development, as well as mechanisms of biomineralization.

A Lineage-Specific Gene Encoding a Major Matrix Protein of the Sea Urchin Embryo Spicule. II. Structure of the Gene and Derived Sequence of the Protein

H. M. Sucov, S. Benson, J. J. Robinson, R. J. Britten, F. Wilt, and E. H. Davidson

Dev. Biol., 120, 507—519, 1987 6-2

A cDNA clone isolated using a polyclonal anti-spicule matrix protein antiserum encodes a prominent 50-kDa spicule matrix protein, SM50. *In situ* hybridization studies confirm that SM50 mRNA is detected only in the skeletogenic primary mesenchyme cells of the sea urchin embryo. The authors have used the cDNA clone to select homologous genomic recombinants and determine the structure of the gene.

The SM50 gene occurs once per haploid genome. It contains a single intron within the 35th codon. Primer extension mapped a unique transcription initiation site 110 nucleotide pairs before the translation start signal. The mRNA is 1895 nucleotides long, excluding a 3′poly(A) sequence, and it contains a single open reading frame 450 codons long. The SM50 mRNA constitutes an estimated 1% of total mRNA in skeletogenic mesenchyme cells. The derived peptide sequence indicated a typical *N*-terminal signal peptide. The protein contains an internal domain quite rich in proline residues and a very basic C-terminal region.

The SM50 protein is the first calcite matrix protein for which the primary sequence is available.

♦Benson and co-workers have recently undertaken several studies to identify components of the mesenchyme cell spicule matrix (*Exp. Cell Res.*, 148, 249, 1983; *J. Cell Biol.*, 102, 1878, 1986). The present reports focus on one of the major protein components of the spicule and with earlier work are important for several reasons. First, they have defined the anatomy of the spicule and the biochemical nature of the components residing within this structure. Second, one of the molecules which may function in the biomineralization process has been characterized and cloned. Third, this spicule matrix protein represents a cell lineage-specific molecule which may be very useful in understanding the differentiation of the mesenchyme cell lineage and expression of cell-type specific genes.

The accompanying papers authenticate the cloned molecule and characterize the genes encoding this molecule. By Northern analysis, the transcript encoding the major spicule matrix protein is shown to accumulate concurrent with differentiation of the mesenchyme cells and by *in situ* hybridizations is shown to be mesenchyme cell specific. The gene is single copy containing a single intron. Furthermore, from the deduced amino acid sequence, the molecule appears to contain acidic 13 amino acid repeats, consistent with an earlier characterization of calcite matrix proteins from three different phyla (Weiner, *Am. Zool.,* 24, 945, 1984). Since this is the first calcite mineral matrix protein sequenced, it should be useful in comparative studies with the many animals which depend on calcite skeletons. *Gary M. Wessel*

Cell Lineage Conversion in the Sea Urchin Embryo

C. A. Ettensohn and D. R. McClay

Dev. Biol., 125, 396—409, 1988 6-3

The conventional view of mesoderm morphogenesis in the sea urchin embryo is that only primary mesenchyme cells (PMCs) produce the larval skeleton while secondary mesenchyme cells (SMCs) remain uninvolved with skeletogenesis. PCMs differ from SMCs in their lineage, developmental fate, and time of ingression into the blastocoel. However, the distinction between these two cell types is clouded both by the fact that they connot be separately identified in certain species and because of previous reports that SMCs can become skeletogenic. The experiments reported here were designed to provide rigorous evidence of SMC conversion and also to further define this phenomenon, using molecular and quantitative techniques.

Removal of PMCs from more than 40 embryos delayed their morphogenesis, but the resulting embryos formed complete skeletons of normal size and shape. Cell marking experiments showed that skeletogenesis in these embryos was due to cell-lineage conversion and not to incomplete removal of PMCs. Time-lapse video microscopy showed that the cells which became skeletogenic appeared to be a subpopulation of SMCs. The earliest detectable event of conversion was the acquisition of cell surface molecules normally unique to PMCs; later events included localization of the converted cells in the ring pattern expected for PMCs. Experiments involving replacement of labeled PMCs into embryos from whom all PMCs had been previously removed showed that the conversion response is a graded one: the number of SMCs converting was inversely proportional to the number of PMCs remaining.

The controlled experiments presented here confirm earlier observations that SMCs can convert to a skeletogenic phenotype normally exhibited only by PMCs. The mechanism for such a transformation is still unclear, as is the issue of whether conversion is due to a change from one differentiated state to another or to the recruitment of undifferentiated cells.

♦Ettensohn and McClay describe an impressive conversion of phenotype in mesenchyme cells of the sea urchin embryo. Historically, mesenchyme cells have been divided into two distinct populations based on their timing of ingression into the blastocoel and their lineage. The first cells to ingress within the blastocoel are the skeletogenic primary mesenchyme cells derived from the 16 cell-staged micromeres. Secondary mesenchyme cells are derived from the veg2 tier of blastomeres at the 64-cell stage (Horstadius, *Biol. Rev.,* 14, 132—179), and they ingress to within the blastocoel during invagination of the archenteron. The fates of this population of cells is less well defined, though subpopulations do give rise to muscle cells and to pigment cells. Evidence also exists that the secondary mesenchyme cells can generate spicules. In fact, Fukushi (*Mar. Biol. Stat. Asamushi,* 11, 21—30, 1962) found that certain embryos depleted of primary mesenchyme cells can still give rise to normal spicules. In the present report, Ettensohn and McClay very carefully extend the observations of Fukushi, and define a cooperative interaction between the primary and secondary mesenchyme cells; secondary cells can convert their phenotype into a primary mesenchyme cell phenotype when the embryo is devoid of primary mesenchyme cells. The form of communication between the primary and secondary cell types is unknown, but implicated are extracellular matrix components synthesized by the primary mesenchyme cells. These findings are also important from an evolutionary standpoint. A primitive sea urchin subclass (Cidaroidea) do not possess the conventional PMC/SMC distinction (Wray and McClay, *Development,* 103, 305—315, 1988). The results from the present study could be interpreted then that the modern (Euechinoidea) species of sea urchin have diverged into a separation of mesenchyme cell lineages (PMC/SMC) of which the SMC have retained the potential of PMC phenotype. *Gary M. Wessel*

---

glp-1 is Required in the Germ Line for Regulation of the Decision between Mitosis and Meiosis in C. elegans

J. Austin and J. Kimble

Cell, 51, 589—599, 1987 6-4

In *Caenorhabditis elegans*, some germ cells undergo mitotic divisions only while others enter meiosis and form gametes. Previous work from this laboratory has shown that in germ cells the decision between mitosis and meiosis is controlled by a somatic cell called the distal tip cell. When the distal tip cell is removed, all germ cells enter meiosis. This study presents a genetic and developmental analysis of mutants in one gene affecting the interaction between the distal tip cell and the germ cells.

Eight recessive mutations were identified which defined the gene germ-line proliferation defective (*glp-1*). These failed to complement each other and all mapped to the same locus on LGIII. In homozygous animals of both sexes harboring the most extreme, presumably null, allele, all germ cells entered meiosis prematurely, after a reduced number of mitoses, and with no death of germ-line cells. Spermatogenesis also occurred early in *glp-1* animals. However, meiosis and gametogenesis appeared normal by both morphological and functional criteria. This phenotype mimicked the effects distal tip cell ablation. Embryos produced by *glp-1* hermaphrodites lacked the anterior bulb of the pharynx, were defective in other aspects of morphogenesis, and were inviable. This embryonic lethality appeared due to a lack of maternal contribution of *glp-1* gene product. Temperature shift experiments suggested *glp-1* activity was required continuously for normal germ-line development. Genetic mosaics were constructed, and their analyses indicated that expression of *glp-1+* in the distal tip cell was neither necessary nor sufficient for normal germ-line cell mitoses.

The authors suggest the wild-type *glp-1* gene product regulates the decision between mitosis and meiosis, and functions in the interaction between germ cells and distal tip cells. They propose that it is either the receptor for the distal tip cell signal or an element involved in the transduction of such a signal.

♦Laser ablation experiments (Kimble and White, *Dev. Biol.*, 81, 208—219, 1981) previously showed that a cell at the tip of the developing gonad called the distal tip cell controls the decision between mitosis and meiosis in the germ line of *C. elegans.* The distal tip cell exerts its influence over a distance on cells that it doesn't contact. This paper identifies one of the components involved in the signaling between distal tip cell and germ line cells, a gene called *glp-1* (germ line proliferation). By making genetic mosaics that were *glp-1* mutant for the distal tip cell and wild type for germ line or *vice versa,* it could be shown that *glp-1* acted in the germ line. The *glp-1* gene product is likely to be involved in receiving a signal from the distal tip cell which suppresses meiosis in mitotically dividing germ cell precursors.

This is an important result since the *glp-1* gene is also apparently used in signaling between early embryonic blast cells to control cell fates. This effect of *glp-1* is characterized in more detail in the accompanying paper by Priess et al. (*Cell,* 51, 601—611, 1987) also abstracted here. *Joseph G. Culotti*

The *glp-1* Locus and Cellular Interactions in Early C. elegans Embryos

J. R. Priess, H. Schnabel, and R. Schnabel

Cell, 51, 601—611, 1987 6-5

In normal development in *Caenorhabditis elegans*, descendants of each of the first two blastomeres, AB and P_1, both differentiate into pharyngeal cells. Previous studies suggested that the differentiation of certain of these descendant cells is dependent on inductive interactions similar to those operating in vertebrate development. The present study was designed to identify genes which function in early cell fate decisions, and resulted in an analysis of mutants that seemed to disturb the interactions between AB and P_1.

A genetic screen was designed to find maternal effect lethal mutations. Four allelic mutations mapped to chromosome III at the same location as that of the independently isolated *glp-1;* the former failed to complement with the latter and so both were deemed to be alleles of the same locus. Progeny of homozygous mothers, but not fathers, were inviable, showing variable abnormalities in elongation, with supernumerary hypodermal cells. Some of the morphogenetic defects appeared to be due to mispositioning of hypodermal cell precursors. These embryos produced only half the number of pharyngeal muscle cells as wild-type embryos, despite a normal number of total body cells. Those cells present were of the same number and type as those normally produced from the P_1 blastomere; when P_1-derived blastomeres were ablated in embryos from homozygous *glp-1* mothers, no pharyngeal muscle cells were produced. In lieu of the absent pharyngeal cells, neuron-like cells were found. Gonadal abnormalities also occurred.

This study permits the conclusion that *glp-1* is a maternal effect mutant. Analyses of the apparent pleiotropic effect of mutations at this locus suggest that the *glp-1* gene acts by promoting cell-cell interactions in wild-type development involving both embryogenesis and gonadogenesis.

♦The fate of blast cell ABa in *C. elegans* was previously shown to be dependent on inductive interactions with P1 or its descendents (Priess and Thomson, *Cell,* 48, 241—250, 1987). In this paper the gene *glp-1* was shown to affect the fate of Aba in a manner that suggested a defect in the inductive interaction with P1. The product of *glp-1* is therefore involved in two types of interactions that control the subsequent fate of one of the cell types involved: (1) the interaction between distal tip cell and germ-line precursors (see accompanying paper by Austin and Kimble, *Cell,* 51, 589—599, 1987) and (2) the interaction between P1 and ABa (this paper). The *glp-1* gene product is maternally inherited (also shown by Austin and Kimble). An interesting possibility which is not mentioned in either paper, is that the *glp-1* gene product may be an excellent candidate for a developmental determinant that is maternally inherited and segregated to specific blast cells (presumably Aba) during early embryonic divisions. *Joseph G. Culotti*

A Cell That Dies during Wild-Type C. elegans Development Can Function as a Neuron in a *ced-3* Mutant

L. Avery and H. R. Horvitz

Cell, 51, 1071—1078, 1987 6-6

Programmed cell death is a common phenomenon, especially in the development of nervous systems. It is possible that cells that die are relics of evolution, which once served a function but are now useless or deleterious. If this were true, it might be possible that prevention of cell death might restore the hidden function of cells which are normally programmed to die. In order to explore this possibility, Avery and Horvitz employed a mutant of *Caenorhabditis elegans, ced-3,* in which cells almost survive that normally die during development.

Experiments were performed which demonstrated that when the pharyngeal neuron M4 is laser ablated during development, the worms starve. This is because the posterior half of the isthmus of the pharynx remains closed, presumably because it lacks its normal innervation by M4. When similar experiments were performed in *ced-3* mutants, partial function of the isthmus occurred often, and full function rarely or never. This partial function appeared to be due to the functioning of MSpaaaaap, a sister cell to M4, which is normally subject to programmed cell death.

The authors conclude that a cell that dies during normal development survives and partially replaces M4 in the *ced-3* mutants. They comment that it is reasonable that variable function is obtained from the unmasking of the possibilities of a cell that is not under any selective pressure. In analogy to the idea of a pseudogene, such pseudocells might be subject to random evolutionary drift.

♦This paper is important because it establishes the concept of a "pseudo cell". In wild type *C. elegans* a large number of specific cells undergo a programmed cell death. In mutants of *ced-3* these cells survive. Apparently a specific spared cell in *ced-3* can only partially and variably replace the function of its sibling as shown by laser ablating the sibling, in *ced-3* animals. Partial and variable replacement may be explained if the cell which normally undergoes a programmed death in the wild type has undergone random drift of its hidden fates — a distinct possibility since it is not under any selective pressure to maintain a particular fate. "Pseudo cells" might play roles in the evolution of cell lineage similar to those proposed for pseudogenes in the evolution of genomes. *Joseph G. Culotti*

A Genetic Pathway for the Specification of the Vulval Cell Lineages of *Caenorhabditis elegans*
E. L. Ferguson, P. W. Sternberg, and H. R. Horvitz
Nature (London), 326, 259—267, 1987 6-7

The complete cell lineage of the nematode *Caenorhabditis elegans* has been determined to delineate mechanisms of cell division and cell fate. Twenty-three genes have been assigned to various steps in a pathway specifying the vulval cell lineages of *C. elegans.* Mutations in most of the genes lead to homeotic transformations in the fates of individual cells. Fifteen genes function in a system involved in the intracellular response to an extracellular signal inducing vulval formation. Mutations in three of five genes required to generate vulval precursor cells cause these cells to adopt fates typical of nonprecursor cells. Mutations in the genes affecting the fates of precursor cells cause certain of these cells to adopt fates characteristic of other precursor cells.

Mutations in some of these genes apparently disrupt or reverse the asymmetry of normally asymmetric cell divisions. If these divisions produce cells with autonomously determined fates, the genes may influence the generation or segregation of or the response to determinants of cell fates that normally are asymmetrically distributed at certain cell divisions. The genes involved in determining precursor cell fates may act in receiving an intercellular signal or in the intracellular response to the signal. Genes involved in the intracellular response to the anchor cell signal may encode membrane or cytoplasmic proteins that function in "second messenger" systems.

♦This paper is among the first to describe the genetic specification of the fate determining mechanism used in specifying an entire organ — the vulva of *C. elegans.* Mutants of more than 20 genes that affect the fates of the vulval precursors are described. The genes can be assigned to specific steps in a developmental pathway for specification of the vulva. Most of the mutants exhibit homeotic transformations, suggesting that specification of specific vulval cell types is accomplished by a series of decisions that distinguish between alternate fates. *Joseph G. Culotti*

Parental DNA Strands Segregate Randomly during Embryonic Development of Caenorhabditis elegans
K. Ito and J. D. McGhee
Cell, 49, 329—336, 1987 6-8

An understanding of whether parental DNA strands segregate together or independently could be important in elucidating lineage-specific gene expression during embryonic development. Previous studies have not shown how DNA strands segregate in the early embryonic divisions of a multicellular organism, when distinct cell lineages are being established. The authors studied DNA strand segregation during development of the nematode *Caenorhabditis elegans,* in which the cell lineage is completely defined.

The fate of gamete DNA was followed in next-generation embryos of *C. elegans* by growing male worms or spermless hermaphrodites on bromodeoxyuridine-containing *Escherichia coli,* in order to label germ-line DNA. Matings produced embryos in which only the DNA strands originating with the gametes contained label, as detected using a fluorescein-labeled monoclonal antibody specific to bromodeoxyuridine. Gamete DNA strands were found to segregate randomly during development.

Random DNA strand segregation implies that, at each embryonic mitosis, the chromosomes have to reengage themselves at random to the mitotic spindle, and have no memory of how they segregate in successive mitoses. The randomness of the process counters developmental models which invoke chromosome imprinting or immortal DNA strands. Embryonic determinants apparently are located in the cytoplasm, although only nuclear determinants which are stably associated with parental DNA strands are ruled out.

♦In theory, one way that developmental lineages could be generated is by somehow marking or imprinting one strand of a chromatid which then gets nonrandomly segregated at each subsequent division to a specific daughter cell. This model is ruled out as a mechanism for generating the well-characterized lineages of the nematode *C. elegans,* since labeled chromatid strands were found to segregate randomly in different individuals. The successful interpretation of these results depended upon the absence of sister chromatid exchange during *C. elegans* development. The results are consistent with the "prevailing view that embryonic determinants are located in the cytoplasm". *Joseph G. Culotti*

Mutations Altering the Structure of Epidermal Growth Factor-like Coding Sequences in the Drosophila *Notch* Locus

M. R. Kelley, S. Kidd, W. A. Deutsch, and M. W. Young

Cell, 51, 539—548, 1987 6-9

The CNS of *Drosophila* develops from progenitor neuroblasts which arise in part of the embryonic ventral ectoderm. The neurogenic protein *Notch* is composed largely of tandemly repeated copies of a sequence resembling epidermal growth factor (EGF). *Notch* protein contains 36 of these EGF-like elements, no two of them identical. DNA from *Ax* and *spl* mutants was cloned and sequenced in an attempt to understand how these mutations affect *Notch* function.

Eight mutations in all were correlated with single amino acid substitutions in EGF-homologous elements of the protein. Genetic analyses of the mutations and comparison of DNA from mutant and wild-type flies suggested that differentiation of function exists among the tandemly repeated EGF-like sequences. The total number of repeats and minor variations form the consensus repeat sequence appeared to have a role in wild-type *Notch* protein function. One group of repeats might be involved in the formation of functional *Notch* dimers or multimeric proteins. Other elements appear to govern interaction of the protein with the product of *Enhancer of split*, another neurogenic locus of *Drosophila*.

EGF-like sequences probably have a common role in providing sites for protein-protein interactions. Differentiation of function reflects different targets for interaction.

♦This is possibly the best example to date of how the molecular characterization of mutant alleles of a fate determining gene product can provide important insights into its molecular mode of action. The *Notch* gene is used by developing neuroblasts in the neurogenic region of the *Drosophila* embryo to inhibit surrounding cells from also becoming neuroblasts. Sequence analysis of the *Notch* gene shows that it encodes a molecule with 36 EGF-like repeats. The molecular characterization of several kinds of mutant alleles show that many of these repeats have independent functions, yet carry out their functions in the context of the whole molecule, i.e., without first being cleaved into single EGF-repeat units. *Joseph G. Culotti*

Indeterminate Cell Lineage of the Zebrafish Embryo

C. B. Kimmel and R. M. Warga

Dev. Biol., 124, 269—280, 1987 6-10

Early cleavages in the zebrafish embryo usually are stereotypes, but some variability is evident. The relation between the site of the plane of the first cleavage and that of the plane of bilateral symmetry is indeterminate, suggesting that cell fate is not determined by cell lineage. The authors studied the relation between lineage and cell fate in the zebrafish by injecting blastomeres of known lineage with a lineage-tracer dye, and following the clonal progeny of injected cells when primary organs had formed and many cell types were differentiating.

Blastomeres arising by the same cleavages in different embryos generated clones in which cell type and position were highly variable. By blastula stages, the cleavages of identified blastomeres exhibited variable patterns. Cell fate could not readily be related to the longitudinal and dorsoventral position of the clone in the gastrula. The clonal progeny of a single blastomere did tend to remain together as a coherent cell cluster through late blastula stages.

In the zebrafish embryo, single blastomeres are able to generate a very diverse array of cell types; cell lineage is indeterminate. It is possible that embryonic cells in the zebrafish bear heritable restrictions, but whether any restrictions are heritable in a cell-intrinsic manner remains to be determined. This fish may partition cells within tissue-specific rather than region-specific boundaries.

♦The relationship between cell lineage and cell fate differs among different organisms. In several lower eukaryotic organisms (e.g., *C. elegans*), the linkage is rigid and cell fate is directly determined by cell lineage. In higher eukaryotic organisms, the relationship is less rigid. In this article, the authors examined this question in the zebrafish. This organism is attractive for such studies because its transparent nature allows direct visualization of internal structures in the living early embryo. Fluorescein-labeled dextran was injected into individual blastomeres and development was allowed to proceed. Subsequent observations indicated that a a rigid relationship between cell lineage and cell fate does not exist in the zebrafish. Individual blastomeres can ultimately differentiate into many cell phenotypes and participate in the histogenesis of several different tissues. *Joel M. Schindler*

Multipotent Precursors Can Give Rise to All Major Cell Types of the Frog Retina

R. Wetts and S. E. Fraser

Science, 239, 1142—1145, 1988 6-11

One means of studying how different cell types are created in correct positions and numbers is to examine the lineages of embryonic cells. The authors have studied cell lineages in the neural retina of the frog by injecting single cells of the optic vesicle with lysinated rhodamine dextran, a fluorescent marker. A prospective lineage analysis was carried out to determine the various cell types formed by individual precursor cells.

Labeled descendants were observed in all three layers of the larval retina following iontophoretic injection of single optic vesicle cells. Different clones consisted of varying combinations of all the major cell types, including glial Muller cells. Some clones were restricted to only one layer.

The wide range of cell types produced by a given precursor raises the question of how the relative nembers of different cell types are regulated. There is evidence that specification may occur via short-range cell-to-cell interactions. Pre-existing neurons in the developing eye vesicle might influence the phenotypes of the cells produced by multipotent precursors.

♦The frog retina is a excellent tissue to study cell lineage relationships because it contains only a few cell types, is experimentally accessible and develops rapidly. For these reasons, the authors explored the cell origins of frog retinal tissue. Lysinated rhodamine dextran was microinjected into

individual frog optic vesicle cells. Observation following further development showed that each precursor cell could give rise to all types of cells in the retina. In addition, it supports the idea that cellular specification occurs late in retinal development, perhaps through specific cell-cell interactions.
Joel M. Schindler

Purification and Characterization of Mouse Hematopoietic Stem Cells

G. J. Spangrude, S. Heimfeld, and I. L. Weissman
Science, 241, 58—62, 1988 6-12

The authors have characterized mouse hemopoietic stem cells using cell-separation methods based on monoclonal antibody binding to cell surface differentiation antigens — which are present on some, but not all cells in a given population or lineage. Multipotentiality of isolated cells was assayed by limit-dilution assays for clonogenic precursors of B lineage cells *in vitro* and T lineage cells *in vivo.*

Stem cells were isolated from murine bone marrow that can proliferate and differentiate into myelomonocytic cells, B cells, or T cells. Thirty cells sufficed to save half of lethally irradiated mice, reconstituting all blood cell types in surviving animals. Thy-1^{lo}Lin$^-$Sca-1^+ marrow cells are a virtually pure population of primitive myeloerythroid stem cells. They probably are the only pluripotent hematopoietic stem cells. Colonies of thymocytes derived from isolated Thy-1^{lo}Lin$^-$Sca-1^+ marrow cells were established after intrathymic injection of as few as four cells.

A similar approach could be used to isolate the human stem cell counterpart if suitable markers are present on human stem cells. An assay for human stem cells is needed. Human fetal liver cells can reconstitute a human lymphoid system in immunodeficient mice, and might suffice for identifying and isolating hemopoietic stem cells. If so, they could be used in marrow transplantation and as targets for gene insertion therapy.

♦Understanding the developmental biology of the hematolymphoid system is predicated on the isolation and characterization of a homogeneous population of self-renewing pluripotent hematopoietic stem cells. Each stem cell must possess the potential to populate all of the known hematolymphoid lineages, and a minimum number of the stem cells must fully reconstitute lethally irradiated animals. Spangrude and colleagues have isolated mouse bone marrow hematopoietic cells which satisfy these two criteria, culminating a search which began more than 30 years ago when it was recognized that lethally irradiated animals could be rescued by injection of bone marrow cells from unirradiated donors. The strategy involved steps of negative and positive selection, using the magnetic separation technique. Stem cells were defined as those which lacked cell surface markers of lineage differentiation (B220, preB, and B lymphocytes; CD4 and CD8, T lymphocytes; Gr-1, granulocytes; Mac-1, myelomonocytic cells) but possessed stem cell antigen-1 (Sa-1) and Thy1.1. The stem cells, designated Thy-1^{lo}Lin^{-} Sca-1^{+}, represented about 0.05% of the total marrow cell population. When injected into lethally irradiated mice, as few as 30 of the stem cells rescued 50% of the animals with repopulation of the T cell, B cell, and myeloid lineages. The Thy-1^{lo}Lin^{-} Sca-1^{+} cells appeared to be pure and to constitute the major if not only population of stem cells in bone marrow, based on their capacity to proliferate and to differentiate into T cells, B cells, and myelomonocytic cells with nearly unit efficiency. Moreover, the number of Thy-1^{lo}Lin^{-}Sca-1^{+} cells necessary for 50% reconstitution of lethally irradiated mice was approximately that predicted by the proportion of these cells in the marrow population. The availability of pure populations of stem cells will not only facilitate the investigation of the early stages of development of the hematolymphoid system but will catalyze progress in clinical therapies involving gene and cell replacement via bone marrow transplantation. It remains to be seen whether human stem cells are similar to those of mouse, an issue which will not be answered until a suitable, e.g., *in vivo,* assay of the human stem cells is developed. *Judith A. K. Harmony*

Hybrid Insulin Genes Reveal a Developmental Lineage for Pancreatic Endocrine Cells and Imply a Relationship with Neurons

S. Alpert, D. Hanrahan, and G. Teitelman
Cell, 53, 295—308, 1988 6-13

The embryologic origin of pancreatic islet cells remains uncertain. Both a neuroectodermal and an endodermal hypothesis have been proposed. The authors used transgenic mice harboring hybrid genes containing the regulatory region of the insulin gene linked with the Tag or PLA protein-coding region in order to assess beta-cell lineage in the developing mouse pancreas.

Insulin appears on embryonic day 12 in the developing mouse pancreas. The transgenic mice first expressed the transgene product 2 d earlier in a few cells of the pancreatic bud. All insulin-producing cells coexpressed the hybrid insulin gene through development and postnatal life. Islet cells containing glucagon, somatostatin, pancreatic polypeptide, and tyrosine hydroxylase — a neuronal enzyme — also expressed the transgene when they first arose. These islet-cell markers were coexpressed during differentiation of the various endocrine cell types. The transgene product appeared transiently in cells of the neural tube and neural crest.

The fact that catecholaminergic neurons and islet cells have phenotypically similar precursor cells indicates that they may be ontogenetically related. Recent work with the preporglucagon gene supports a relationship between islet cells and neurons.

♦The embryological lineage of endocrine cells in the pancreas is still unknown. One theory hypothesizes a neuroectodermal origin. The study reported here provides some new and important insight into this old question. The authors utilize a hybrid gene construct containing the regulatory information of the insulin gene and sequences encoding either the SV40 T antigen or human placental lactogen as reporter genes. Unlike cell ablation studies to distinguish lineage, these studies monitor coexpression of the hybrid gene with known endogenous gene products. The results indicate that the hybrid gene is expressed in all four pancreatic endocrine phenotypes and transiently in cells of some neural tissue. These observations support the hypothesis that pancreatic endocrine cells arise from a common precursor. They also suggest that such a precursor might be of neuroectodermal origin, or also participate in the development of neural tissue. *Joel M. Schindler*

Cytodifferentiation — Cell and Tissue Specific Gene Expression and Maintenance 7

INTRODUCTION

Once cells respond to certain cytoplasmic signals, express a certain set of early developmentally regulated genes, migrate and interact with their neighbors, and become determined to a certain cell lineage, they "finally" differentiate. In so doing, they mold tissue composition and the final array of cell phenotypes in the developed organism.

Whether the developed organism contains only a few terminally differentiated cell phenotypes (e.g., slime mold), or many (e.g., man), certain crucial signals must exist to assure the proper expression of the appropriate genetic repertoire. Because of these signals, a fully functional system, like the nervous system or the immune system, can become operational despite being comprised of many related cell phenotypes.

Molecular genetic techniques have allowed the extensive dissection of several structural features of specific genes that are necessary for tissue-specific expression. Engineering fusion genes and constructing transgenic animals have greatly enhanced the ability to explore the nature of tissue-specific genetic signals.

Several organ systems, specifically the nervous system, the immune system, and gonadal tissue, have received a great deal of attention. These demonstrate a high level of cellular and physiological complexity. Sex determination is obviously of vital importance since any abnormalities could impact on the ability to produce future generations.

Once differentiated, certain cell phenotypes require factors to remain functional. Proto-oncogenes, several of which encode growth factors, may therefore play a role in either the establishment or the maintenance of differentiated phenotypes.

The papers included in this chapter demonstrate examples of cell- and tissue-specific genes and gene products and their coordinated regulation. Neuronal development, immune development, and sex determination are highlighted along with the technologies that are employed to facilitate such investigations.

Structural Characterization of *Dictyostelium discoideum* Prespore-Specific Gene D19 and Its Product, Cell Surface Glycoprotein PsA

A. E. Early, J. G. Williams, H. E. Meyer, S. B. Por, E. Smith, K. L. Williams, and A. A. Gooley

Mol. Cell. Biol., 8, 3458—3466, 1988 7-1

The *Dictyostelium discoideum* cell surface antigen PsA is a glycoprotein appearing in the multicellular stage, soon after tip formation, and is selectively expressed on prespore cells. PsA is the best-characterized cell type-specific surface molecule of the multicellular stage. The authors determined the entire sequence of the D19 gene, which encodes an mRNA sequence highly enriched in prespore over prestalk cells in the slug stage. The gene was shown to encode the psA protein.

An 81-amino acid portion of N-terminal sequence from immunoaffinity-purified PsA protein was identical to the predicted sequence of the D19 gene. Several short repeat elements are present close to the C terminus. Unequal crossing-over within these elements may explain the size polymorphism noted in PsA protein isolated from different strains of *D. discoideum.* The repeats resemble the C-terminal region of the *D. discoideum* cell adhesion molecule. When the PsA gene was marked by inserting an oligonucleotide encoding an epitope of human c-*myc* protein, a construct of this gene with 990 base pairs of 5′-flanking region direct temporally and spatially correct mRNA accumulation. The marked PsA protein was correctly localized on the cell surface.

It now is possible to use transformation to localize the signals required for prespore-specific gene expression, and to study possible functions of the protein.

♦The PsA prespore-specific surface antigen and its corresponding gene have been isolated and characterized. The PsA antigen was purified by immunoaffinity on columns derivatized with the monoclonal antibody MUD 1, which specifically recognizes this antigen. The PsA appears to be anchored in the plasma membrane via a glycosyl-phosphatidylinositol moiety. There are no useable sites for asparagine-linked glycosylation. This confirms other reports that the oligosaccharides on this protein are in O linkage. The PsA gene was marked with an oligonucleotide encoding an epitope of the human *c-myc* protein. This construction, including the 990 bp 5′ upstream region was transfected into *Dictyostelium* cells. This 5′ sequence appears to have all the regulatory regions necessary for the proper developmentally regulated transcription of the D19 gene. *Stephen Alexander*

Antibodies to a Fusion Protein Identify a cDNA Clone Encoding msp130, a Primary Mesenchyme-Specific Cell Surface Protein of the Sea Urchin Embryo

D. S. Leaf, J. A. Anstrom, J. E. Chin, M. A. Harkey, R. M. Showman, and R. A. Raff

Dev. Biol., 121, 29—40, 1987 7-2

Primary mesenchyme cells are determined very early in sea urchin embryogenesis. They exhibit a largely autonomous set of cellular activities during differentiation. The authors found that the primary mesenchyme cell-specific cDNA clone 18C6 encodes a 130-kDa protein, designated msp130 (mesenchyme surface protein, 130-kDa).

The cDNA clone has been partly sequenced, and an open reading frame identified. Part of this is expressed as a beta-galactosidase fusion protein in *Escherichia coli*, and antibodies to the fusion protein were generated. The antibodies recognize msp130, and this is the same 130-kDa protein recognized by the primary mesenchyme-specific monoclonal antibody B2C2. The latter recognizes a post-translational modification of the protein. The transcript encoding msp130 was first detected in premesenchyme blastula embryos. The transcript accumulated to a significant degree after primary mesenchyme cell ingression, and msp130 protein was first detectable soon after ingression.

The msp130 protein may have a role in cell migration, aggregation, and/or spicule formation. The protein has been implicated in both calcium uptake and in cell migration/pattern formation processes.

♦Several laboratories recently have generated interesting antibody probes to mesenchyme cell-specific, surface glycoproteins in the sea urchin embryo. These probes have been useful for several studies on spiculogenesis, cell lineage specific differentiation, and cell surface interactions. Leaf et al. have identified a cDNA clone which appears to unify many of these antibody studies. In this report, a clone termed msp130 is defined and characterized by antibodies to a fusion peptide which recognizes the same molecule as several of the other mesenchyme cell specific glycoproteins. This clone represents a 3.8-kb transcript and a 130-kDa protein. The transcript accumulates just prior to ingression of the mesenchyme cells and is shown to be mesenchyme cell specific. The molecule is a cell surface glycoprotein and appears to be involved in some calcium ion uptake events during spiculogenesis (Carson et al., *Cell,* 41, 639—648, 1985). In conjunction with several other reports on mesenchyme cells this year (Benson et al., *Dev. Biol.*, 120, 499; Ettensohn and McClay, *Dev. Biol.*, 125, 396; Decker et al., *Development,* 101, 297), it is clear that we have learned a great deal about this interesting cell type in the sea urchin embryo. *Gary M. Wessel*

Spec3: Embryonic Expression of a Sea Urchin Gene Whose Product is Involved in Ectodermal Ciliogenesis

E. D. Eldon, L. M. Angerer, R. C. Angerer, and W. H. Klein

Genes Dev., 1, 1280—1292, 1987 7-3

The authors describe a sea urchin embryo mRNA with an unusual temporal and spatial accumulation pattern. The regulation of Spec3 *Strongylocentrotus purpuratus* ectoderm) expression suggests that it is related to ciliogenesis. A cDNA clone specifying Spec3 mRNA originally was isolated from a differential screen of a pluteus cDNA library, using radiolabeled probes derived from pluteus ectoderm vs. endoderm/mesoderm fractions.

Spec3 mRNA is present at low levels in unfertilized eggs and accumulates rapidly during cleavage. Levels then fall abruptly and rise again during the prism and pluteus stages. A similar pattern is described for ectodermally expressed beta-tubulin mRNAs. Spec3 message at first accumulated in all blastomeres and later was restricted to ectoderm. By pluteus, grains were concentrated in the ciliary band. Deciliation and regeneration of cilia in gastrula-stage embryos led to a four- to fivefold increase in Spec3 mRNA levels.

The single gene in the haploid genome that encodes Spec3 mRNA contains three exons encoding an open reading frame for a hydrophobic protein of 21.6 kDa. The carboxy-terminal part of the protein contains two long hydrophobic stretches separated by short hydrophilic regions. Probably the Spec3 protein contains 1 alpha-helical domains that either span the lipid bilayer or are associated with another hydrophobic environment.

The Spec3 message appears to encode a protein related to ectodermal ciliogenesis; it may have other tubulin-associated functions as well. Identical regulatory mechanisms may control the expression of the Spec3 and beta-tubulin genes.

♦Ciliogenesis is a distinct and well-studied event in sea urchin embryogenesis. A single cilium forms on the apical side of all blastomeres (except in some species, the most vegetal blastomeres lack them) when the cleavage cycle slows and the embryo reaches the hatching blastula stage. This occurs at the eighth to ninth cleavage. The cilia on ectoderm cells are for swimming and sensory functions, while those on presumptive endoderm cells become luminal cilia of the gut after gastrulation. There are over 150 different structural proteins that make up the cilium. Some of these components have been the subject of recent investigations, in particular the alpha and beta tubulins (Harlow and Nemer, *Genes Dev.*, 1, 147—160, 1987; Gong and Brandhorst, *Mol. Cell. Biol.*, 7, 4233—4246, 1987). However, most ciliary proteins are uncharacterized in terms of their structure, function, and biogenesis. The present article describes the properties of Spec 3 mRNA which appears to encode a novel ciliary protein. The accumulation of the Spec3 message closely resembles that of an ectoderm specific beta tubulin mRNA identified by Nemer and colleagues (Harlow and Nemer, 1987, op. cit.) both temporally and spatially, and with respect to decilation and regeneration. The latter observation provides the best evidence, albeit circumstantial, that the Spec3 protein is associated with ectodermal cilia. From the putative protein sequence it appears that the Spec 3 protein may be part of the ciliary membrane. If true, this would be significant since membrane proteins specifically associated with cilia are not well characterized. Recent unpublished results (Eldon, Decker, Brandhorst, and Klein) using an antibody against the N terminus of the Spec3 protein show that Spec3 is present on the outside surface of ectodermal but not endodermal cilia. It is also present at high levels in the golgi which lie just underneath the base of the cilium in each cell. It is not known whether Spec3 is a specialized component of sea urchin cilia or is more widespread. The ectodermal location of the Spec3 mRNA and protein suggest that its function may be associated with the external environment. *William H. Klein*

Correct Cell-Type-Specific Expression of a Fusion Gene Injected into Sea Urchin Eggs

B. R. Hough-Evans, R. R. Franks, R. A. Cameron, R. J. Britten, and E. H. Davidson

Dev. Biol., 121, 576—579, 1987 7-4

When a fusion gene construct containing 5′ flanking sequences of the CyIIIa cytoskeletal actin gene of *Strongylocentrotus purpuratus*, ligated to coding sequences for bacterial chloramphenicol acetyltransferase (CAT), is injected into unfertilized sea urchin eggs, embryos begin synthesizing CAT enzyme at the early blastula stage. The mRNA produced by the CyIIIa actin gene is from the outset confined to cells giving rise to aboral ectoderm, and the gene continues to be expressed only in these cells.

After injecting the fusion gene construct into unfertilized sea urchin eggs, early pluteus-stage embryos were analyzed by *in situ* hybridization, using a ^{3}H-RNA antisense probe for CAT mRNA. Autoradiography showed that CAT mRNA was present only over aboral ectoderm cells, and not over any recognizable regions of gut or oral ectoderm in the same sections.

These findings indicate the cell type-specific localization of CAT mRNA in sea urchin embryos. Spatial control of CyIIIa gene expression probably occurs through interaction with *trans*-regulators that are stored, synthesized, or presented in the aboral ectoderm progenitor cells during cleavage.

♦This article presents the first evidence for proper spatial expression of an exogenously introduced gene into sea urchin embryos. Davidson and colleagues have been studying the differential expression of the actin gene family in sea urchins for a number of years. One actin gene, CyIIIa, is activated early in embryogenesis and is expressed exclusively in aboral ectoderm cells of the embryo and larvae. In earlier papers, these workers showed that linear DNA can be injected into unfertilized eggs of *S. purpuratus* and maintained in a concatenated, amplified form during embryogenesis. Microinjections of fusion genes with 5′ flanking regions of the CyIIIa actin gene and the bacterial CAT gene demonstrated that CAT activity appeared at the early blastula stage, the time when the endogenous CyIIIa actin gene is normally activated. In the present report, an *in situ* hybridization assay with a CAT probe shows that eggs microinjected with a CyIIIa-CAT fusion gene result in early plutei where CAT transcripts can be detected only in patches of aboral ectoderm cells. No signal above background is observed in gut, oral ectoderm, or mesenchyme cells. The authors suggest that the partial labeling pattern observed in the embryos is due to the mosaic distribution of the microinjected DNA. The concatenated DNA is thought to be incorporated into a given nucleus at different stages of cleavage and thus expression from the injected DNA will differ from one embryo to the next depending on which cell-lineage the DNA becomes associated with, and the time in which the event occurs. The authors argue that because no patterns are seen that are not subsets of aboral ectoderm, proper spatial expression must be occurring. The work is of particular importance because it shows that *cis* regulatory elements conferring cell-type specificity will function properly when present on microinjected plasmid molecules introduced into sea urchin eggs and that cell-type specific expression from exogenous genes can be analyzed in sea urchin embryos by *in situ* hybridization. *William H. Klein*

DNA Synthesis and the Control of Embryonic Gene Expression in C. elegans

L. G. Edgar and J. D. McGhee
Cell, 53, 589—599, 1988 7-5

The nematode *Caenorhabditis elegans* is useful in studying how DNA synthesis influences lineage-specific gene expression during development, because of its highly defined cell lineages. The gut lineage, which is clonally descended from only one cell of the 8-cell embryo, is especially convenient to study. The authors used the DNA polymerase inhibitor aphidicolin to determine how inhibition of DNA synthesis influences gene expression in the developing gut of *C. elegans.*

Microspectrofluorimetric studies of DNA synthesis showed that aphidicolin inhibited synthesis nearly instantly and completely. Marker expression — both of gut granules and a carboxylesterase — was not controlled by reaching the normal DNA:cytoplasm ratio, by counting the normal number of rounds of DNA synthesis, or by simple lengthening of the cell cycle. Expression of both markers required a brief period of DNA synthesis in the first cell cycle after clonal establishment of the gut.

After an early period of DNA synthesis, the timing of gene expression is independent of both DNA synthesis and cytokinesis. Markers receive permission for later expression during a limited period of DNA synthesis just after clonal establishment of the gut, and further DNA synthesis is not required. It may be during this critical period that cytoplasmic determinants migrate to the nucleus and take their place at specific sites on the chromosomes.

♦This is an important paper because it reaffirms Holzer's quantal cell cycle model (Holzer et al., *Rev. Biophys.*, 8, 532—557, 1975) concerning the expression of tissue specific markers. Edgar and McGhee show that tissue-specific expression in the gut of *C. elegans* depends on a brief period of DNA synthesis probably during the first cell cycle after the tissue has been clonally established. Tissue-specific expression does not depend on DNA synthesis after this. Possibly during this critical period of DNA synthesis, cytoplasmic determinants migrate into the nucleus, bind to specific sites, and later influence the expression of tissue-specific gene products. These results also suggest that gene expression during intestinal development does not depend on DNA:cytoplasm ratio nor by passage through a "counted" number of rounds of DNA synthesis. *Joseph G. Culotti*

Identification and Intracellular Localization of the *unc-22* Gene Product of *Caenorhabditis elegans*

D. G. Moerman, G. M. Benian, R. J. Barstead, L. A. Schriefer, and R. H. Waterston

Genes Dev., 2, 93—105, 1988 7-6

Among the first genes isolated in *Caenorhabditis elegans* by Tc1 tagging was the *unc*-22 locus, a frequent target of the element. The *unc*-22 locus is one of more than 25 genes involved in normal muscle formation in *C. elegans.* Mutant alleles of the locus lead to impaired movement and muscle disorganization. The authors used DNA clones from the *unc*-22 region to study the molecular organization of the gene and the nature of its product.

The *unc*-22 gene extends over more than 20 kb of genomic sequence and produces a 14-kb transcript. A polyclonal antibody against an *Escherichia coli* beta-galactosidase-*unc*-22 fusion protein recognized a polypeptide in nematode extracts that labeled the muscle A-band on indirect IF microscopy. In mutants the large protein was absent or severely reduced in amount. A truncated polypeptide was identified in one instance.

The *unc*-22 gene product may interact with myosin to regulate its function. Further study of this and other genes may help in understanding thick filament assembly and function.

♦This paper represents one of the first examples of transposon tagging to clone a gene, and use of the cloned gene to make a specific antibody to its product in order to examine its localization *in situ*. This approach should ultimately define all of the components used in muscle development, function, and stabilization in *C. elegans*. The nematode genes may help to identify and examine the function of similar muscle-specific gene products in vertebrate muscle. *Joseph G. Culotti*

A Genetic Analysis of the Sex-Determining Gene, *tra*-1, in the Nematode *Caenorhabditis elegans*

J. Hodgkin

Genes Dev., 1, 731—745, 1987 7-7

In *Caenorhabditis elegans*, animals with a single X chromosome become males, while those which are XX develop into hermaphrodites, essentially modified females. Mutations in at least seven autosomal genes can alter this process. For one of them, *tra-1*, complete sexual transformations have been demonstrated. Loss-of-function alleles cause XX animals to become fertile males, while gain-of-function alleles cause XO animals to become fertile females. The other six genes may serve mainly to mediate between the primary sex-determining signal, X chromosome dosage, and *tra-1*.

The authors conducted an extensive mutational analysis of the *tra-1* gene. The most severe mutations transformed XX animals into fertile males, but had little effect on XO animals. There were some abnormalities in XO gonadal development. Weaker *tra-1* alleles led to incomplete masculinization of XX animals and various intersexual phenotypes. All gain-of-function alleles transformed XX animals into fertile females. The two functionally different types of allele appeared to map to different locations on the large gene.

Activity of the *tra-1* gene is feminizing and specific to XX animals. In XO animals, the gene assists normal male somatic gonad development and promotes spermatogenesis. The *tra-1* gene has most attributes of a "complex" gene, in the sense of some *Drosophila* loci.

♦*Tra-1* is a major switch gene controlling sexual phenotype in *C. elegans*. This paper describes the phenotypes of 65 mutant alleles of *tra-1*; some loss of function alleles transform XX animals into males while other gain-of-function alleles transform XO animals into females or hermaphrodites. The various roles of *tra-1* in the soma, in the somatic gonad, and in the germline differ. Altogether the results suggest that this is a complex gene with multiple (and sometimes genetically separable) functions in different cells. The molecular analysis of this gene should prove particularly useful in unraveling these functions. It will be especially interesting to know if there are counterparts to *tra-1* in vertebrates, and if so, how these might be related to elements of the sexual pathway controlled by the testis-determining factor gene in humans (see Page et al., *Cell*, 51, 1091—1104, 1987 and review by Hodgkin, *Nature (London)*, 331, 300—301, 1988). *Joseph G. Culotti*

Detection *in situ* of Genomic Regulatory Elements in *Drosophila*
C. J. O'Kane and W. J. Gehring
Proc. Natl. Acad. Sci. U.S.A., 84, 9123—9127, 1987 7-8

Random generation of operon fusions in prokaryotes involves integrating a promoterless reporter gene, whose expression is readily detected, and assaying at several positions in a target genome so that it comes under control of randomly selected chromosomal promoters. Genes then may be isolated and characterized solely by knowing or postulating their pattern of expression, without having to screen for mutant phenotypes. In eukaryotes such as *Drosophila*, the reported gene normally would have to be the first gene in the fusion transcript, and an active promoter would be needed.

The authors used a fusion of the reporter gene *lacZ* of *Escherichia coli*, which codes for beta-galactosidase, to the P-element promoter of the *Drosophila* genome. Of 49 transformed fly lines, about 70% exhibited some type of spatially regulated expression of *lacZ* in embryos. Many of these expressed *lacZ* specifically in the nervous system.

The P-*lacZ* fusion gene is an efficient means of recovering elements that may regulate gene expression in *Drosophila*, and of generating a wide range of cell type-specific markers. This approach should be applicable to other eukaryotic organisms.

♦This paper describes a novel approach for the screening of transcriptional regulatory elements residing in the *Drosophila* genome. The principle of the approach is related to one previously used in prokaryotes, whereby a promoterless reporter gene is integrated at random sites in the bacterial genome. In a certain proportion of the cases, fusion with endogenous promoters occurs, as indicated by activity of the reporter gene. The present set of experiments was designed to detect the presence of tissue-specific enhancer elements. The bacterial beta-galactosidase (lac-Z) gene was fused to a weak and constitutive promoter that functions in all tissues (the P element promoter) in a construct that also carried a wild type copy of the *rosy* eye color gene. This construct was transformed into the germ line of *rosy* flies. Embryos from transformed lines (as determined by the wild-type eye color phenotype) were screened for tissue-specific expression of the beta-galactosidase gene using a whole-mount staining procedure. Normally the P element promoter is too weak to generate enough beta-galactosidase to be detected by this relatively insensitive procedure. About 20% of the transformed lines belonged to this class. The remaining 80% of the lines stained relatively strongly, presumably because the P element promoter was stimulated by a nearby enhancer element. Of the strongly staining line, a surprisingly high proportion (37 of 49, or about 70%) showed some sort of spatially restricted embryonic expression. Specific expression in the nervous system was the most frequent pattern (13 of 37 cases). Other transformant lines showed specific expression in a restricted part of the gut, in the epidermis, in the malpighian tubules, in the muscles, and in the amnioserosa.

The tissue-specific expression of the reporter gene is almost certainly due to the influence of enhancer-like, tissue-specific regulatory elements present near the point of integration. Therefore, one might expect to find a nearby gene that is expressed with similar tissue specificity. Moreover, because the transforming DNA includes the gene coding for the dominant *rosy* eye color marker, it should be possible to obtain deletions (e.g., with X-rays) in neighboring genes in one generation. Thus, this procedure offers a surprisingly easy molecular access to genes that are specifically expressed in any of a large number of tissues and cell types of the fly. This approach may prove to be particularly fruitful and informative, and in principle also be applicable to other organisms. *Marcelo Jacobs-Lorena*

Interacting Genes that Affect Microtubule Function: the *nc2* Allele of the *haywire* Locus Fails to Complement Mutations in the Testis-Specific β-Tubulin Gene of *Drosophila*

C. L. Regan and M. T. Fuller

Genes Dev., 2, 82—92, 1988 7-9

Complex microtubular arrays mediate several functions in normal cells. Many aspects of microtubular functional regulation — and structural interactions between microtubules and other organelles — may be mediated by accessory proteins. A mutation that fails to complement certain alleles of the testis-specific β2-tubulin gene *(B2t)* of *Drosophila* maps to a separate locus, *haywire.* Second-site noncomplementing mutations such as *hay*^{nc2} and *B2t* alleles could serve to identify genes that encode products participating in the same functions, or interacting in the same structure.

Genetic interaction between *hay*^{nc2} and *B2t* required the presence of the mutant *hay* gene product, suggesting a structural interaction between this product and β2-tubulin. A deficiency for the *hay* region complemented the same alleles of *B2t* that *hay*^{nc2} failed to complement. The latter is a recessive male sterile mutation in a genetic environment that is wild type at the *B2t* locus. Homozygous males exhibited defects in meiosis, flagellar elongation, and nuclear shaping, all microtubule-based processes. Homozygous *hay*^{nc2} females had reduced fertility.

These findings are consistent with a role for the *haywire* gene product in microtubular function. Analysis of second-site noncomplementing mutations is a promising approach for identifying genes encoding products which interact to build complex protein arrays.

♦The *haywire*^{nc2} (*hay*^{nc2}) mutation was isolated in a screen for new alleles of a testis-specific beta-tubulin (*B2t*) gene. In such a screen, the majority of the noncomplementing mutants obtained represent new alleles in the beta-tubulin locus itself. However, an exceptional mutation (*hay*^{nc2}) was obtained that failed to complement the *B2t*^{3} allele but that maps to a separate site elsewhere in the third chromosome. Thus, *B2t*^{3}/+; *hay*^{nc2}/+ flies are defective while *B2t*^{3}/+; +/+ or +/+; *hay*^{nc2}/+ flies are wild type. The *hay*^{nc2} mutation was found to map to none of the known tubulin loci of *Drosophila*. It is likely that the *hay*^{nc2} gene product interacts in some way with the male-specific beta-tubulin protein. In support of this hypothesis is the finding that a deficiency for the *hay* region complements the same alleles of *B2t* that *hay*^{nc2} fails to complement. Moreover, *hay*^{nc2} complements certain alleles of the beta-tubulin gene while it fails to complement other alleles of the same gene. It is possible that the *hay* gene product is a microtubule-associated protein. Consistent with this hypothesis is the fact that homozygous *hay*^{nc2} male mutants are sterile and show defects in the three major microtubule-based processes in which the testis-specific beta2-tubulin participates: meiosis, flagellar elongation, and nuclear shaping. Moreover, homozygous mutant females have severely reduced fertility in addition to generalized temperature-sensitive semilethal effects.

A likely scenario for this mutation is that the *hay*^{nc2} product binds tubulin protein molecules and forms a complex which is defective in a function required after binding. In heterozygous *B2t* mutants, the already reduced level of wild-type beta-tubulin protein would be further reduced by complexing with the defective *hay* product, thus reducing functional beta-tubulin levels to below a minimum threshold concentration.

The main significance of the paper is that it represents one of only a few examples in higher eukaryotes where noncomplementing second-site mutations have been used to identify genes whose products are likely to interact. In principle this approach could also be used to identify components of other structures which are composed of multiprotein arrays. *Marcelo Jacobs-Lorena*

Regulation of Sexual Differentiation in D. melanogaster via Alternative Splicing of RNA from the *transformer* Gene

R. T. Boggs, P. Gregor, S. Idriss, J. M. Belote, and M. McKeown
Cell, 50, 739—747, 1987 7-10

The *transformer* locus is one of the genes that controls somatic sex determination in *Drosophila melanogaster*. Although *tra* is only necessary for female development, in addition to a 1-kb female specific transcript, this locus also produces a 1.2-kb transcript that is present in both sexes. To understand the function and regulation of these two transcripts, genomic and cDNA *tra* clones were sequenced.

One of the introns in the *tra* transcript is processed in a sex-specific manner. The nonsex-specific RNA uses a 3′ splice acceptor site at position 185, while the female-specific transcript uses a 3′ splice acceptor site at position 360. The nonsex-specific transcript contains many stop codons in all three reading frames. The female-specific transcript has an open reading frame that could encode a peptide of 22 kDa. The nonsex-specific transcript appears to be unable to encode a product.

This work demonstrates that the *tra* locus is sexually regulated through alternative splicing. Only the female-specific transcript is capable of encoding the *tra* product.

The Control of Alternative Splicing at Genes Regulating Sexual Differentiation in D. melanogaster

R. N. Nagoshi, M. McKeown, K. C. Burtis, J. M. Belote, and B. S. Baker

Cell, 53, 229—236, 1988 7-11

The primary determinant of sex in *Drosophila* is the ratio of sex chromosomes to autosomes. Genetic experiments suggest that somatic sex determination is further controlled by a series of genes: *Sxl→tra, tra-2→dsx,ix→* terminal sex differentiation. The regulation of the *tra* and the *dsx* locus were examined at the molecular level in this study.

The pattern of transcription from the wild-type *tra* gene was examined in chromosomally female flies mutant at *Sx1, tra-2* or *dsx.* Wild-type *Sx1* function was necessary for the female-specific *tra* transcript to be synthesized. Therefore, *Sx1* acts upstream of *tra.* Neither *tra-2* nor *dsx* had any effect on *tra* expression. Similar experiments demonstrated that *Sx1, tra,* and *tra-2* were necessary for female-specific *dsx* function.

The molecular data generated in this study is consistent with a regulatory hierarchy: *Sx1→tra, tra-2→dsx,ix.* This is also consistent with the genetic data.

Ectopic Expression of the Female *transformer* Gene Product Leads to Female Differentiation of Chromosomally Male Drosophila

M. McKeown, J. M. Belote, and R. T. Boggs

Cell, 53, 887—895, 1988 7-12

These authors have previously shown that sex-specific expression of *tra* mRNA is controlled at the level of splicing. Ectopic expression of the two *tra* mRNAs was used to examine the regulatory hierarchy of the genes controlling somatic sexual differentiation.

A cDNA copy of the female-specific *tra* mRNA was fused to an *hsp70* promoter and introduced into the genome of wild-type flies by germ-line transformation. All XY flies that carried this construct had female development, demonstrating that expression of the *tra* female-specific mRNA is sufficient to induce female specific somatic differentiation. These results are consistent with a linear model for the somatic sexual differentiation hierarchy. Expression of the hs-tra-female construct overcame the sex transformation of *Sx1* mutants, indicating that *Sx1* acts upstream of *tra*. All the other somatic sexual differentiation genes were required to be wild type for female differentiation in the presence of tra-female transcription. Therefore, all of these genes are downstream of *tra* in the regulatory cascade. In a line of flies deleted for *tra* but carrying hs-tra-female, all flies develop as females, indicating that the nonsex-specific *tra* transcript is without function in females.

These results suggest that the genes that control somatic sexual differentiation are regulated as a linear cascade. These data further suggest that the nonsex-specific *tra* transcript is without function.

♦The decision between development along the male or female pathways in *Drosophila* is determined by the X chromosome to autosome ratio (X:A ratio). Somatic sexual differentiation is mediated by a small number of genes and genetic studies suggest that information is transferred in the following hierarchical order: X:A ratio to *Sex-lethal* (*Sxl*) to transformer (*tra*), transformer-2 (*tra-2*) to *doublesex* (*dsx*), *intersex* (*ix*) to terminal differentiation (where separated by a coma, ambiguity exists about the order of the two genes). Experiments reported in these papers confirmed and extended this model.

The paper by Boggs et al. concerns the regulation of *tra* gene expression. A puzzling observation was that despite the fact that *tra* has no known function in males (this gene is only required for female development), *tra* is equally transcribed in males and females. Thus, control of expression of the *tra* gene does not occur at the transcriptional level. A major finding by Boggs et al. was that regulation actually occurs at the level of RNA splicing. In males, the primary *tra* transcript yields an RNA containing no reading frame of significant length, and is presumed to have no function. In females the *tra* precursor RNA is spliced to yield an RNA with a long open reading frame, predicting a 22-kDa protein with a very basic central domain that could be involved in nucleic acid binding.

Nagoshi et al. used the sex-specific splicing pattern of *tra* and *dsx* to verify the interactions among the different genes that had been deduced from genetic experiments. To illustrate a typical experiment, the pattern of *tra* expression in chromosomally female (X:X) *Sxl* or *dsx* mutants was analyzed on Northern blots. Recall that in the regulatory cascade *Sxl* is presumed to be situated "upstream" and *dsx* "downstream" of *tra*. In *Sxl* mutants only the male *tra* transcript could be detected, even though the flies are chromosomal females. In *dsx* mutants the female pattern of *tra* transcription is observed, even though these flies develop as males. This experiment shows that *Sxl* is situated above *tra* in the regulatory cascade and *dsx* below *tra*.

McKeown et al. have constructed a gene that ubiquitously expresses the female-specific transcript (female cDNA under the control of the heat shock promoter) and shows that this gene is sufficient to direct female development in chromosomally male (X:Y) individuals. As expected, these individuals also display a "female pattern" of *dsx* transcripts. Moreover, the same gene can also "correct the phenotype" of *tra* mutants. A similar construct expressing the "male" *tra* RNA has no effect in either males or females.

Collectively, these experiments confirmed the hierarchical order of action of the sex determining genes. Moreover, these molecular experiments eliminated several alternative models, including models invoking branched pathways, that had not been previously excluded by genetic evidence.

While many examples of alternative splicing exist, in only very few cases has it been shown to be involved in developmental or tissue-specific regulation of gene function. For this reason, these studies are important. These findings also raise a possible molecular explanation for the interesting observation that *Sxl* is involved in the perpetuation ("cellular memory") of the X:A signal by a mechanism of autoregulation. The proposal is that *Sxl* is not only regulated at the level of splicing but also that regulation occurs by action of its own protein product. Because the *Sxl* gene has been cloned, the hypothesis is clearly testable. *Marcelo Jacobs-Lorena*

Molecular Genetics of the *single-minded* Locus: A Gene Involved in the Development of the Drosophila Nervous System

J. B. Thomas, S. T. Crews, and C. S. Goodman

Cell, 52, 133—141, 1988 7-13

A search for mutations that disrupt embryonic neural development in *Drosophila* uncovered the *sim* locus. This locus is the same as the previously identified lethal complementation group *1 (3) S8.*

Homozygous *sim* mutants are late embryonic lethals. At 11 h, these embryos were missing the cells derived from the midline strip of cells, including 2 MP1 progeny, the ventral unpaired median neurons, and the MECs. The *sim* gene was found to be single copy. Two major transcripts were detected, a 3.0-kb transcript that peaked at 6 h and then disappeared and a 3.5-kb transcript that peaked at 6 to 9 h and was expressed throughout the rest of embryogenesis at a low level. *In situ* hybridization to whole embryos demonstrated that *sim* expression can be detected just prior to gastrulation in a two-cell-wide strip of cells along the border of the presumptive mesoderm. By 6 h, hybridization can be detected over the entire strip of midline cells, and by 11 h, hybridization can be detected along the midline of the developing nervous system.

The phenotype of *sim* mutants and the temporal and spatial pattern of expression of *sim* suggest that this gene plays an important role in the development of the cells along the midline of the developing central nervous system in *Drosophila* embryos.

The Drosophila *single-minded* Gene Encodes a Nuclear Protein with Sequence Similarity to the *per* Gene Product
S. T. Crews, J. B. Thomas, and C. S. Goodman
Cell, 52, 143—151, 1988 7-14

Mutation of the *single-minded (sim)* gene results in the loss of precursor cells that lie along the midline of the central nervous system of the *Drosophila* embryo. The sequence of the *sim* product was determined, and antisera were raised to increase understanding of its function.

The largest *sim* cDNA clone was sequenced and revealed one open reading frame. The corresponding amino acid sequence was deduced and compared to sequences in the Doolittle and NBRF data bases. The only significant similarity was to the product of the *Drosophila per* locus. The *sim* protein sequence was hydrophilic and the carboxy-terminal half of the protein contained many repetitive sequences. Between amino acids 364—404, the sequence Ala-Ala Gln was repeated imperfectly 14 times. A series of diverse homopolymeric stretches were also observed. Antibodies were raised to a recombinant *sim* protein and used to stain embryos. The antiserum stained the nuclei of a strip of cells at the ventral midline in gastrulating embryos.

The *sim* protein appears to be expressed in those cells that are lost in *sim* mutants. This suggests that *sim* expression is required for the proper development of these cells. As this protein is nuclear, it may act by regulating other genes involved in the development of the neuroepithelium of the midline of the embryo.

♦Gastrulation in *Drosophila* starts 3 h postfertilization, immediately after cellularization of the blastoderm is completed. At this point the embryo consists of monolayer of cells arranged as an ovoid. The future mesoderm is generated by the invagination at the ventral midline of an anterior-posterior stripe of cells approximately one dozen cells wide. The cells that originally bordered the sides of the invaginated stripe are brought together at the ventral midline. Later cells belonging to this second narrow stripe also invaginate to give rise to a small set of specialized neuronal and non-neuronal (glial?) cells of the central nervous system.

These papers deal with the morphological, genetic, and molecular characterization of the *single-minded* (*sim*) locus. Homozygous *sim*⁻ embryos specifically lack the specialized set of neuronal and non-neuronal cells referred to above. The gene was cloned, a *sim* cDNA was sequenced, and antibodies were obtained to a recombinant *sim* protein. *In situ* hybridization with the cloned probe revealed that before blastoderm *sim* is expressed only in a narrow one- or two-cell wide strip bordering both sides of the presumptive mesoderm. After invagination of the mesoderm, hybridization is observed only to cells lying along the ventral midline. In older embryos (11 h of development), only a specialized subset of the central nervous system cells appears to express the gene. A similar pattern was observed when embryos were stained with *sim* antibody. Thus, there is an excellent correlation between the cells that express the gene and the cells that are absent in the mutant embryos. In addition, antibody staining showed that the protein is nuclear. Sequence comparison revealed sequence similarity with the *period* gene (see Report 6 in this series) that controls circadian rhythms in *Drosophila*. The significance of this similarity is not clear at present.

It is striking that expression of this neurogenic locus is restricted to such a sharply delineated narrow stripe of cells as early as the cellular blastoderm, at a time when all cells are morphologically identical. As is the case with other developmentally important genes, expression occurs well in advance of overt morphological differentiation. In all likelihood, *sim* expression is connected to the maternal and early zygotic genes that direct differentiation along the dorsal-ventral axis of the embryo. *Marcelo Jacobs-Lorena*

Expression and Function of the Segmentation Gene *fushi tarazu* During *Drosophila* Neurogenesis

C. Q. Doe, Y. Hiromi, W. J. Gehring, and C. S. Goodman
Science, 239, 170—175, 1988 7-15

The *fushi tarazu (ftz)* gene is transiently expressed in the central nervous system (CNS) during *Drosophila* embryogenesis. To examine the role of *ftz* during embryogenesis, its pattern of expression in the CNS was studied in wild-type flies. Flies that do not express *ftz* in the CNS were created and examined.

Antisera to the *ftz* product and beta-gal assays in flies carrying *ftz-lacZ* fusions were used to monitor *ftz* expression in embryos. Expression was detected in the CNS after 5 h. By 8 h, *ftz* was expressed in approximately 30 nuclei per hemisegment. By 12 h, *ftz* expression was no longer detectable. To test for *ftz* function during neurogenesis, *ftz* expression in the CNS was selectively turned off by promoter mutation, and this construct *(ftzK)* was transformed into a *ftz*-background. Although *en* expression was unchanged, both *Ubx* and *eve* expression was altered in different ways in different neurons in *ftzK* embryos. Although aCC, pCC, MP1, dMP2, vMP2, and RP1 *ftz*-expressing neurons had unaltered morphology in *ftzK* flies, the RP2 *ftz*-expressing neuron appeared to be transformed toward RP1.

This transformation suggests that loss of *ftz* expression in the CNS causes a change in the identity of at least one neuron, RP2. This supports the hypothesis that *ftz* has a regulatory role in the determination of neuronal fate during development.

Control of Neuronal Fate by the *Drosophila* Segmentation Gene *even-skipped*

C. Q. Doe, D. Smouse, and C. S. Goodman

Nature (London), 333, 376—378, 1988 7-16

Many of the *Drosophila* segmentation genes are also expressed in the developing central nervous system. For example, the *eve* protein can be detected at 5 h and then persists throughout embryogenesis in the central nervous system. To selectively inactivate central nervous system expression of *eve,* a temperature-sensitive mutant of *eve* was used.

Embryos that were homozygous for *eve-ts* were raised at 18° C and then shifted to 30° C during neurogenesis. This specifically inactivated *eve* function in the central nervous system. Neurons aCC and RP2, which normally express *eve,* had aberrant axon morphology.

Many other segmentation genes of *Drosophila* are expressed in the central nervous system. These genes may also be involved in the generation of neuronal diversity.

♦During early embryogenesis the segmentation and homeotic genes play a key role in subdividing the embryo into segments and in establishing the identities of these segments. Many of these genes are also expressed in the nervous system of later embryos but their function in neural development remains unknown. One of the difficulties in determining their role in the nervous system is that mutants in these genes disrupt earlier steps of embryogenesis to such an extent that the nervous system does not form properly. This barrier was circumvented in these two studies in two elegant ways: (1) expression of the *fushi tarazu* (*ftz*) gene was selectively eliminated only from the nervous system by deleting the "neurogenic element" from its promoter, and (2) function of the *even-skipped* (*eve*) gene was suppressed specifically during neurogenesis by appropriate temperature shifts of embryos carrying temperature-sensitive (ts) *eve* mutation. In wild-type embryos *ftz* is transiently expressed in only 30 of 250 neurons of each hemisegment while *eve* is expressed in 16 of these 250 neurons. Suppressing the activity of either of the two genes during neurogenesis does not affect the overall organization of the nervous system, as determined by staining with an antibody that recognizes all axons. Moreover, the specific neurons that normally express *ftz* and *eve* are viable and the cell bodies occupy their normal position even in the absence of *ftz* and *eve* activity. However, the axons of some of these neurons follow aberrant pathways.

This phenotype is not simple to interpret since not all neurons that normally express *ftz* or *eve* are disrupted by elimination of the corresponding gene activity. Moreover, a complex relationship exists in the expression of *Ultrabithorax, eve,* or *ftz* in these neurons that does not relate directly to phenotype.

In summary, the axonal pathway and possibly axon identity is disrupted in certain of the neurons that express *ftz* and/or *eve,* when the activity of one of these genes is supressed. It should be noted that these studies were restricted to examining the morphology and hnot the function of the neurons. However, they represent an excellent first step in the very difficult task of establishing the role that homebox-containing genes play in the developing nervous system. *Marcelo Jacobs-Lorena*

Probes for Rare mRNAs Reveal Distributed Cell Subsets in Canary Brain

D. F. Clayton, M. E. Huecas, E. Y. Sinclair-Thompson, K. L. Nastiuk, and F. Nottebohm

Neuron, 1, 249—261, 1988 7-17

The vertebrate brain contains more than 30,000 different species of polyadenylated RNAs. In an attempt to understand the organization of these RNAs within the brain, mRNAs that were differentially expressed between the canary forebrain and the rest of the brain were identified and studied.

Forebrain cDNA was inserted into pGEM-3, to allow synthesis of high-specific-activity antisense riboprobes. Plus/minus colony hybridization was used to detect clones of RNAs present at greater than 10^{-4} in the forebrain population. Hybridization of cDNAs to purified plasmid DNAs from insert-containing clones was used to detect mRNAs present at 10^{-4} to 10^{-6} abundance. Subtractive cDNA probe enrichment was used to detect mRNAs present at less than 10^{-6} abundance in forebrain. Seven specific differentially expressed polyA+ RNAs were identified in this screen. Quantitative filter hybridization was used to estimate abundances, which ranged from 10^{-4} to 4×10^{-7}. Cellular distribution was determined by optimized *in situ* hybridization, replacing the RNase step with a stringent wash at 65° C in 11 m*M* sodium, which resulted in greater sensitivity with lower background. Clone pCF-2 was very abundant in the cerebellum, particularly in the Purkinje cells, clone pCF-17 was abundant in the lobus paraolfactorius, clone pCF-29 was highly specific for 4 to 8% of the cells of the anterior hyperstriatum accessorium and posterior neostriatum, clone pCF-68 was abundant in large neuronal nuclei located in cell aggregates, clone pCF-82 reacted with glia-like cells, clone pCF-125 reacted with a cell that occurred occasionally in several brain regions, and clone pCF-37 was highly specific for the layer of cells that line the ventricles.

These results suggest that thousands of low abundance RNAs may be differentially expressed in the brain. The data are consistent with quantitative regulation of many genes across many different cell sets in overlapping anatomic distributions.

♦The nervous system consists of an extraordinarily diverse array of cell types. During development, divergent sets of neurons differentiate in a region-specific manner. Different brain regions are responsible for various behaviors which presumably depend upon the types of neurons and connections formed by that region. One approach for understanding the regional complexity of the brain is to look for differential gene expression in the brain regions.

Here, Clayton et al. identify several rare RNA sequences that are predominant in the forebrain region of the canary brain. Since the forebrain is the region containing much of the song control circuitry, their aim is to find probes for brain cells involved in this behavior. This paper is interesting and highly useful because it describes a new technique for detecting rare message in the brain. Since those components involved in differential gene expression tend to be rare, this approach provides the opportunity of more sensitive detection than previously possible. The authors have identified several rare mRNAs that are prevalent in large neurons (e.g., Purkinje cells and cells with large nuclei in the forebrain) and not detectable in non-neuronal cells. These results suggest that specific genes are present in many cells types with overlapping anatomical distributions, rather than in specific subregions. The cell types sharing a gene may have related function (e.g., shared neurotransmitters) or may be involved in the same specific circuits. The results indicate that tremendous cellular diversity could be generated by a relatively small number of genes if they were expressed in a combinatorial fashion. *Marianne Bronner-Fraser*

Distinct Neurotrophic Factors from Skeletal Muscle and the Central Nervous System Interact Synergistically to Support the Survival of Cultured Embryonic Spinal Motor Neurons

U. Dohrmann, D. Edgar, and H. Thoenen

Dev. Biol., 124, 145—152, 1987 7-18

Embryonic chick motor neurons do not survive in culture without a soluble muscle-derived factor. Even in the presence of this factor, less than 10% of these cells survive. As these isolated motor neurons have lost spinal cord contact, the influence of spinal cord cells and factors on motor neurons in culture was examined.

Isolated motor neurons survived in heterogenous spinal cell cultures without the addition of exogenous factors. Survival was increased by the addition of the previously defined muscle cell-derived factor. Conditioned medium from the spinal cell culture was sufficient to provide the motor neuron survival promoting factor. Optimal concentrations of spinal cell conditioned medium acted synergistically with the muscle cell-derived extract to promote neuron survival and increased the total survival above that seen with either factor individually. Spinal cell conditioned medium did not promote the growth of sympathetic neurons. Brain-derived neurotrophic factor, nerve growth factor, and acidic and basic fibroblast growth factor were unable to promote motor neuron survival in culture.

Spinal cells produce a soluble factor that promotes the survival of embryonic chick motor neurons. This factor acts synergistically with the previously identified muscle-derived factor.

♦After differentiation, many neurons require exogenous factors for their survival. In the case of motor neurons, it is clear that both peripheral targets and afferent central nervous system connections are important for survival *in situ.* In attempts to identify pertinent survival factors, motor neurons typically are placed in culture. One factor which promotes motor neuron survival *in vitro* is apparently derived from the peripheral targets: a soluble factor from muscle prolongs viability of a small percentage of motor neurons in culture. However, the majority of neurons do not survive under these conditions.

In the present study, Dohrmann et al. examined the effects of a spinal cord-derived factor on motor neuron survival in culture. They find that a soluble factor promotes survival of some motor neurons and that this factor is distinct form previously identified survival factors. Most interesting, the combination of muscle and spinal cord-derived factors has a synergistic effect on motor neuron survival. This suggests that the spinal cord and muscle-derived factors are distinct and that they can act cooperatively. These are interesting findings in that they point to a logical source from which to purify a motor neuron survival factor. Furthermore, they support the notion that both the immediate environment of the motor neurons somas (i.e., the spinal cord) and connections with the target site (i.e., the muscle) are important for the continued viability of motor neurons. Thus, survival of a population of neurons depends upon both interactions with the central nervous system and with the periphery. *Marianne Bronner-Fraser*

Brain-Derived Neurotrophic Factor Prevents Neuronal Death *in vivo*

M. M. Hofer and Y. -A. Barde

Nature, 331, 261—262, 1988 7-19

The survival of developing neurons is thought to depend on the availability of specific neurotrophic proteins. The only protein for which the neuron survival-promoting effect has been demonstrated *in vivo* is nerve growth factor (NGF). The *in vivo* neuron survival-promoting effects of brain-derived neurotrophic factor (BDNF) were investigated in quail embryos.

From day 3 to day 7 of embryogenesis, 10 µg of NGF or 1 µg of BDNF were injected into quail embryos. The embryos were killed and their neurons counted on day 8. Injections of NGF increased survival of neurons in the dorsal root ganglion. Injections of BDNF produced similar results. Injections of BDNF also increased neuron survival in the nodose ganglion. Injections of NGF had no effect in this region.

BDNF increased the survival of developing neurons in two regions of the quail embryonic brain. BDNF may play an important role in the development of sensory pathways.

♦The peripheral nervous system is derived from the neural crest and the ectodermal placodes. A question of great interest is what factors influence neuronal development. After their differentiation, many neurons depend upon exogenous factors for survival, some of which may be acquired by means of their synaptic connections. For example, neural crest-derived neurons are known to depend upon nerve growth factor (NGF) for continued viability. Although a variety of factors have been purified and are thought to be important for neuronal survival, NGF is the only one previously shown to increase neuronal viability *in vivo.*

In this report, Hofer and Barde have tested the effects of brain-derived neurotrophic factor (BDNF), a putative survival factor purified from the brain, on developing neurons in the peripheral nervous system of the quail. The effects of BDNF were compared with NGF. Both BDNF and NGF were administered at daily intervals to quail embryos and the number of surviving neurons were counted. BDNF significantly increased the number of neurons surviving in both neural crest-derived dorsal root ganglia and in the nodose ganglia, whose neurons are derived from the ectodermal placodes. NGF also increased neuronal survival in the dorsal root ganglia, but not in the nodose ganglia. The significance of these results is twofold. First, they demonstrate that BDNF is a neuronal survival factor which increases neuronal survival *in vivo.* This makes BDNF the only factor other than NGF which has been shown to have significant biological effects *in situ.* Second, BDNF is a broader spectrum neuronal survival factor than NGF, since it effects both neural crest- and ectodermal placode-derived neurons. This paper shows that this rare growth factor does effect neuronal development and maintenance and suggests that exogenous BDNF can prevent naturally occurring cell death in a variety of neurons. Furthermore, the findings support the notion that trophic factors can be obtained from both central and peripheral connections. *Marianne Bronner-Fraser*

Expression of a Single Transfected cDNA Converts Fibroblasts to Myoblasts

R. L. Davis, H. Weintraub, and A. B. Lassar

Cell, 51, 987—1000, 1987 7-20

When the embryonic mouse fibroblast cell line C3H10T1/2 (10T1/2) is treated with 5-azacytidine, a demethylating agent, myogenic clones develop. Furthermore, when DNA from such clones is transfected into the parent cells, myoblasts are obtained at frequencies suggesting that only one or a few loci are involved. The work described here was performed in order to identify those loci which transformed fibroblasts into determined myoblasts.

Using subtracted cDNA hybridization between 10T1/2 and 2 myogenic cell lines, probes were prepared which were used to screen a λgt10 myoblast/myotube cDNA library. Three clones were selected, one of which converted 10T1/2 cells into stable myoblasts, when transfected. The transfectants showed myosin-positive, multinucleate syncytia. When transformed into eight other types of cell lines, myosin heavy chain synthesis and cell fusion were detected in seven. Northern analysis showed expression of RNA transcribed from the transfected gene (MyoD1) occurred in neonatal and adult skeletal muscle, but in no other tissue tested. Sequence analysis showed one open reading frame which resembled zinc-finger proteins. This open reading frame was strongly similar to the portion of c-*myc* necessary for its *in vitro* transforming ability and homologous to the portion of the *Drosophila achaete-scute* complex which is expressed in association with neurogenesis.

These experiments have resulted in the isolation of a gene, MyoD1, the expression of which transforms fibroblasts into myoblasts. The subtracted hybridization procedure used to accomplish this may be useful for identifying other sequences that regulate differentiation.

♦The number of expressed genes necessary to produce a specific cell phenotype is unknown, and undoubtedly differs from phenotype to phenotype and organism to organism. In this article, the authors demonstrate that the expression of a single transfected cDNA can elicit myogenic determination from mouse 10T1/2 fibroblast cells. Whether the protein encoded by this cDNA is the first in a cascade of genes necessary for myogenic determination, or the final protein necessary for such determination, is not known but is currently under investigation. This article directly demonstrates that transfection studies can be used to explore the cell-specific function of individual genes. In this way genes which express cell-specific proteins can be identified as well as genes that directly regulate cell-specific differentiation. *Joel M. Schindler*

Diabetes and Tolerance in Transgenic Mice Expressing Class II MHC Molecules in Pancreatic Beta Cells

D. Lo, L. C. Burkly, G. Widera, C. Cowing, R. A. Flavell, R. D. Palmiter, and R. L. Brinster

Cell, 53, 159—168, 1988 7-21

A common belief is that destruction of pancreatic β cells, resulting in insulin dependent diabetes, is an autoimmune response. This has led to speculations that the β cells initiate this response by presenting cell surface antigens to immune cells. Evidence exists that β cells are sometimes capable of expressing class II major histocompatibility complex (MHC) molecules, which are associated with the presentation of antigens to T lymphocytes. This may induce an immune response that results in destruction of β cells, but an alternative possibility is that MHC expression in β cells occurs only secondary to an inflammatory response. The experiments presented here were designed to distinguish between these possibilities.

Transgenic mice were created that express the class II MHC heterodimer $E_{\alpha}^{d}/E_{\beta}^{b}$ (I-E^{b}) on β cells under the control of a rat insulin enhancer and promoter. These mice strongly expressed the transgene in islet, but not exocrine pancreatic tissue and also in the proximal tubules of the kidney. Hyperglycemia developed in these animals by 3 to 5 weeks of age, and profound polydipsia/polyuria by 2 months. The diabetes was clearly insulin dependent. Histological and immunohistochemical analyses failed to detect infiltrating lymphocytes or other inflammatory cells or antibody deposits in the pancreas. Furthermore, tolerance to the transgene (I-E^{b}) molecule was induced, although the diabetic mice were clearly immunocompetent.

Apparently, expression of class II MHC molecules by β cells causes diabetes without involvement of immune mechanisms. This may be due to direct action of large amounts of class II molecules within β cells themselves. The unexpected induction of tolerance to the transgene I-E in the absence of thymic expression of I-E may provide new insight into the mechanism of tolerance acquisition.

Diabetes in Transgenic Mice Resulting from Over-Expression of Class I Histocompatibility Molecules in Pancreatic β Cells

J. Allison, I. L. Campbell, G. Morahan, T. E. Mandel, L. C. Harrison, and J. F. A. P. Miller

Nature (London), 333, 529—533, 1988 7-22

While evidence from both human and animal studies suggests that insulin dependent diabetes mellitus (IDDM) is an autoimmune disease, no direct evidence exists for the assumption that immune cells directly cause the disease by destroying pancreatic β cells. The present work was undertaken to test the hypothesis that expression of a nonself class I major histocompatibility complex (MHC) molecule on the surface of β cells would result in IDDM secondary to a T cell immune response.

Transgenic mice were made to express the H-2K^b MHC class I molecule specifically in β cells, under the direction of the rat insulin promoter. All such mice developed IDDM, irrespective of whether they were syngeneic or allogenic to the transgene. Histological techniques demonstrated abnormally organized pancreatic islets, with depletion of β cells, but no lymphocytic invasion either of islets or of other tissues. Control experiments showed that the development of IDDM in transgenic mice was always associated with expression of H-2K^b, and did not merely depend on reducing insulin gene expression by transgene competition for transcription factors. Other experiments were designed to eliminate the possibility that expression of H-2K^b from the transgene was in any way different from that of the endogenous H-2K^b and thus provocative of an immune response. These involved the use of thymectomized mice, demonstrably immune incompetent. These thymectomized transgenic mice developed IDDM indistinguishable from that of intact transgenic mice.

The authors conclude that overexpression of class I MHC molecules in β cells causes IDDM without T cell involvement. They speculate that since MHC class I molecules are known to bind peptides, overexpression might interfere with insulin processing or secretion.

♦The association of insulin-dependent diabetes mellitus (IDDM) with certain DR-DQ haplotypes of the class II (HLA-D) major histocompatibility complex (MHC) gene region and the pathophysiology of the disease has led to the conclusion that IDDM is an autoimmune disease with the immune response directed toward pancreatic antigens. The contribution from different mechanisms to the establishment of tolerance to self antigens, for example, anergy, suppression, and/or clonal deletion, has not been established for antigens which exist in extrathymic tissues and is not fully understood for antigens which are expressed in the thymus, the organ of T cell ontogeny. Therefore, the basis by which tolerance is broken in autoimmune diseases is unknown. One proposal holds that pancreatic antigens are tolerated because the molecules which are required to present antigen to lymphocytes are not expressed by pancreatic cells, as in the case of class II MHC gene products, or are expressed in low abundance, as in the case of class I MHC gene products. Tolerance abrogation in IDDM is proposed to occur subsequent to the aberrant expression of pancreatic MHC genes, an idea which is based on the increased abundance of the gene products on cells of the diabetic pancreas. To test this hypothesis, two groups of investigators have targeted MHC genes for pancreas-specific expression in transgenic mice by using the insulin promotor to direct MHC gene expression. Lo and colleagues produced transgenic mice which expressed class II I-E^b on beta cells; Allison and co-workers produced mice which expressed class I H-2K^b on their beta cells. In both cases, the beta cells of the transgenic mice degenerated, the levels of insulin were reduced, and the mice developed diabetes. Remarkably, the pathophysiology of the disease did not resemble that of IDDM since no landmarks of an immunological contribution to the tissue destruction were evident. Whereas expression of the MHC transgene by pancreatic beta cells was directly associated with the onset of diabetes, the basis of its effect was not clarified by these investigations. Both groups speculated that MHC gene expression somehow interferes with the normal function of the beta cells. Allison et al. pointed out the possibility that intracellular MHC molecules might interfere with normal insulin processing and trafficking by binding proinsulin or insulin itself. Also, unexpectedly, the transgenic animals were tolerant of incompatible MHC transgenes. The cause of tolerance induction was not established. These investigations raise two intriguing possibilities: (1) that IDDM is not an autoimmune disorder, but is initiated by aberrant MHC gene expression in beta cells with a consequent defect in insulin production, and (2) that tolerance to extrathymic antigens is established by mechanisms different from that for thymic antigens or requires antigen communication between extrathymic and thymic tissues. *Judith A. K. Harmony*

Cell-Autonomous Action of the Testis-Determining Gene: Sertoli Cells are Exclusively XY in XX<—>XY Chimaeric Mouse Testes

P. S. Burgoyne, M. Buehr, P. Koopman, J. Rossant, and A. McLaren
Development, 102, 443—450, 1988 7-23

It is widely believed that testes development from an indifferent gonad required the activity of a diffusible molecule, controlled by the Y chromosome, and capable of affecting both XX and XY cells. This view is supported by the fact that chimeric mouse embryos usually become fertile males and only rarely become hermaphrodites or females. The concept of a diffusible testis organizer has been questioned, however, since embryonic XX gonad tissue cocultured with developing testes does not form testis cords. It is possible that testes in XX<—>XY chimeras could be due to ovarian regression caused by Mullerian-inhibiting substance, and not due to the actions of a diffusible testis organizer on XX cells. The present study investigates this question by analyzing the distribution of XX and XY cells in XX<—>XY chimeric testes.

Cells of XX<—>XY chimeras were analyzed by isoenzyme identification and by *in situ* hybridization of Y chromosome-specific satellite DNA probes or parent-specific probes. Isoenzyme analysis showed that despite the presence of 48 to 56% XX cells in nontesticular tissues and in tunica albuginea, Leydig and Sertoli cells were predominantly XY. *In situ* hybridization to Y chromosome probes resulted in all Sertoli cells scoring positive. In two male chimeras, XX parent-specific probes failed to hybridize to any Sertoli cells or germ cells. Control experiments showed that these cells bound the XX-parent specific probe and that Sertoli cells of XY<—>XY chimeras bound the XY-parent specific probe.

These data show that in XY<—>XY chimeras the Sertoli cells are exclusively XY. The authors suggest that their results contradict the notion that the testis-determining gene *(Tdy)* encodes a diffusible factor to which both XX and XY cells can respond. Instead, they believe that their data support the notion that Sertoli cell differentiation depends on cell-autonomous Y chromosome activity.

♦This paper presents evidence that contradicts the long standing view that the testis-determining factor encoded for by the *Tdy* gene acts via a diffusible molecule to which XX cells respond. In XX<—>XY chimeric mouse testes analyzed by enzyme as well as DNA markers, XX cells were found to contribute to all components of the testis except the Sertoli cells. These cells appeared to be exclusively XY, suggesting that Sertoli cell differentiation is the result of cell-autonomous activity of the *Tdy* gene and that subsequent steps in testis differentiation may be a consequence of Sertoli cell activity. However, contradictory data exist which indicate that transdifferentiation of XX follicle cells into Sertoli cells can occur if, for example, XX fetal mouse ovaries are grafted to male kidneys. These are cases where Sertoli cells form without any Y-chromosomal involvement at all. The authors reconcile these data as not being representative of the normal process of testis development, and therefore, do not discount the fact that *Tdy* may act cell autonomously to trigger Sertoli cell differentiation.
Terry Magnuson

The Sex-Determining Region of the Human Y Chromosome Encodes a Finger Protein

D. C. Page, R. Mosher, E. M. Simpson, E. M. C. Fisher, G. Mardon, J. Pollack, B. McGillivray, A. de la Chapelle, and L. G. Brown
Cell, 51, 1091—1104, 1987 7-24

Because the presence of a Y chromosome in mammals causes an embryo to develop as a male, it is presumed that at least one gene on the Y chromosome directly or indirectly determines all male characteristics. Candidates for this "testis-determining factor" (TDF, or Tdy in mouse) have included the male-specific histocompatibility antigen H-Y and "Bkm" DNA satellite DNA present on the Y chromosome. Here, Page and coauthors set out to clone TDF by deletion mapping. This approach relied on the existence of sex-determining Y chromatin in "XY females."

A sequence of 140 kb of Y chromosome DNA was identified which seemed necessary and sufficient to determine maleness. A portion of this sequence includes 1.3 kb which appears to define an exon of the TDF. This sequence is highly conserved in mammals and even in birds. Similar sequences exist on human X chromosomes. Because it hybridizes to a bovine cDNA library it appears to be transcribed. The homologous region in mouse maps to the sex-determining region of the Y chromosome, lending support to the idea that this sequence encodes the TDF. This sequence appears to encode a zinc finger protein like transcription factors IIIA and Sp1, and which is believed to define a nucleic acid binding domain.

If indeed the sequence described here encodes the TDF, these results suggest that the first step in mammalian male determination is cell autonomous. The authors postulate that TDF is a factor regulating the transcription of genes involved in early events of sex determination.

♦The availability of various structural abnormalities of the human Y chromosome has made it possible to construct a deletion map by hybridizing with Y-DNA probes, and subsequently, to position the testis determining gene (*TDF*) on this map. These genetic data place the *TDF* gene in a small interval that measures 140 kb. By chromosome walking, 230 kb covering this region was cloned, and it appears that only one gene is contained within the area. Sequences located in this region have been found to map specifically to the sex determination region of the mouse Y chromosome, and similar sequences have also been found on the human X chromosome. The predicted amino acid sequence from the open reading frame in this area indicates a protein with multiple finger domains, similar to those found in frog transcription factor IIIA. If this putative protein is the testis determining factor it would be important to characterize the genes whose expression it regulates. Questions that are raised by the authors concern specificity of stage and cell-type expression, and whether TDF is involved not only in establishment but also maintenance of the sexually determined stage. In addition, would such a protein have functions other than gonadal sex determination, and what is the function of the X-homologue. *Terry Magnuson*

Three-Dimensional Structure of an Oncogene Protein: Catalytic Domain of Human C-H-*ras* p21

A. M. de Vos, L. Tong, M. V. Milburn, P. M. Matias, J. Jancarik, S. Noguchi, S. Nishimura, K. Miura, E. Ohtsuka, and S.-H. Kim
Science, 239, 888—893, 1988 7-25

Besides being the most commonly found oncogenes in human tumors, *ras* genes appear to be involved in proliferation, differentiation, and other processes. It has been suggested that, like G proteins, they are signal transducers that may begin these processes. It is known that they bind and hydrolyze GTP and that they have sequence similarity to the GTP-binding domains of G proteins. The work in this report was undertaken to explore the three-dimensional structure of *ras* proteins to help explain the biochemical roles of both activated and normal proteins.

Residues 1 to 171 of *ras* protein encode the "catalytic domain," which has GTPase activity. This portion of c-H-ras, was crystallized, and the crystal structure at 2.7-Å resolution was then determined.

The *ras* protein consists of a six-stranded β sheet, four α helices, and nine connecting loops. An alternating stretch of β-α-β-α-β toward the C terminus has similar topology to a motif previously characterized as a nucleotide-binding domain. Four of the loops form a pocket for GDP binding, and the side chain of Phe^{21} is nearly perpendicular to the guanine. Those point mutations which activate the transforming ability of the protein are located within three of these four loops, and may affect GDP or GTP binding or hydrolysis. A candidate for the GTPase active site is in loop L1, within a glycine-rich stretch; amino acid substitutions at Gly^{12} activate *ras*. Adjacent to this loop is, in viral *ras* proteins, Thr^{59}, the autophosphorylation site. The recognition site for a neutralizing monoclonal antibody to *ras* is in loop L4, opposite the GDP-binding site. Presumably, antibody binding there affects L1 indirectly.

These data help explain the functions of both normal and active *ras* and identify regions responsible for key functions. This knowledge may be useful in drug design.

♦This article has generated considerable interest because it is the first report of the three-dimensional structure of an oncogene product. The ras oncogene family is the most prevalent type of oncogene to occur in human tumors, and has been implicated in the normal processes of cell proliferation and differentiation. Mammalian ras proteins, comprised of 187 or 188 amino acids with Mr ~21 kDa, bind guanine nucleotides GTP and GDP and the proto-oncogenic forms catalyze GTP hydrolysis. They also share sequence homologies with other guanine nucleotide binding proteins such as elongation factor (EF)-Tu and the G proteins which transduce signals between an activated membrane receptor and an intracellular effector. It is generally accepted that ras proteins, like G proteins, function as signal transducers. The three-dimensional structure of residues 1-171 encompassing the catalytic domain, reported here to a resolution of 2.7 Å, reveals the pattern of the polypeptide backbone, the GTP binding site, and putative GTPase catalytic site, and some information about certain amino acid side chains but not information about the precise location of each side chain or of interside chain interactions. Nevertheless, the results provide a framework for interpreting the biochemical and genetic data which implicate regions in the ras protein as being indispensable for biological activity. C-H-ras (1-171) is folded into a single unit consisting of a six-stranded β sheet, four α-helices, and nine connecting loops. The alternating α/β topology is similar to that of all nucleotide-binding domains, although the location of the N terminal β strand in C-H-ras is unique in this comparison. GDP was bound in a packed formed by four loops with phe^{28} oriented nearly perpendicular to the guanine. Activation of the transforming capacity of ras *in vivo* and *in vitro* is associated with point mutations within three of these four loops and particularly with substitution of glycine at position 12, consistent with the hypothesis that activation alters the binding of guanine nucleotides and/or GTPase activity. Greater resolution will be required for a detailed comparison of the three-dimensional structures of ras and other guanine nucleotide-binding proteins such as EF-Tu. Kim and colleagues indicate that they have extended the resolution to 2.25 Å, so the necessary refinements in the structural analysis will soon be forthcoming. *Judith A. K. Harmony*

Expression of the Proto-Oncogene *int*-1 Is Restricted to Specific Neural Cells in the Developing Mouse Embryo

D. G. Wilkinson, J. A. Bailes, and A. P. McMahon

Cell, 50, 79—88, 1987 7-26

Besides their association with tumors, the expression of many oncogenes has been shown to be developmentally regulated, both temporally and spatially. Recent work has suggested that the protooncogene *int-1* is expressed during mouse embryogenesis as well as in mammary tumors. The experiments reported here were undertaken to localize *int-1* expression in the mouse embryo.

In situ hybridization was performed, using an *int-1* probe, against longitudinal and cross sections of embryos of different ages. Accumulation of *int-1* RNA was detected only in particular areas of the neural plate and its derivatives, beginning in the 9-day-old embryo. Initially, *int-1* expression was detected rostrally, in the anterior head folds, and as the neural tube develops and closes, *int-1* expression proceeded caudally. Following closure of the neural tube, *int-1* expression was detected in parts of the brain ventricles and spinal cord, midbrain and diencephalon, and neuroepithelium at the midbrain-hindbrain junction.

These results confirm and extend previous whole embryo RNA blot studies. They provide further evidence for the involvement of *int-1* in central nervous system development, but the significance of the details of the developmental expression of this oncogene is at present unclear.

Expression of the Proto-Oncogene *int*-1 Is Restricted to Postmeiotic Male Germ Cells and the Neural Tube of Mid-Gestational Embryos

G. M. Shackleford and H. E. Varmus

Cell, 50, 89—95, 1987 7-27

The *int-1* protooncogene is active in mammary carcinomas induced by mouse mammary tumor virus. Previous work from this laboratory showed that the *int-1* protooncogene is also expressed in the normal adult mouse, but only in testes, and also during mouse embryogenesis. Because this pattern of expression seemed puzzling, the experiments reported here were undertaken to describe the expression in more detail, in attempts to understand the function of *int-1*.

Sensitive hybridization techniques were employed to survey 22 different adult mouse tissues for *int-1* gene expression, but only testis gave positive results. Germ-cell fractionation, temporal expression studies, and spermatogenesis mutants were all used to analyze *int-1* expression in testis. All techniques gave similar results: *int-1* is expressed only in haploid, round spermatids in the process of spermiogenesis. Other experiments involved dissection of mouse embryos and analyses of the RNA of various parts. These studies showed *int-1* accumulation only in neural tube (diencephalon, mesencephalon, metencephalon, myencephalon, and spinal cord, but not telencephalon).

This work demonstrates expression of *int-1* in male germ cells undergoing spermiogenesis and in certain embryonic neural tube cells. Such diverse developmental expression, combined with the association of *int-1* expression in mammary tumorigenesis, might be explained if the *int-1* protein is a secreted ligand which reacts with membrane-bound receptors of different cells, affecting each differently. An alternative explanation might be that the *int-1* protein is differentially cleaved or the *int-1* transcript is differentially spliced, yielding different products with different functions.

♦The proto-oncogene *int*-1 thus far has been found to be active only in mammary carcinomas that have been induced by the mouse mammary tumor virus. Insertion of the provirus is responsible for activating transcription of the proto-oncogene. Normal expression of *int*-1 does occur, but as described in these reports, it is restricted in the adult to the postmeiotic germ cells undergoing differentiation from round spermatids into mature spermatozoa, and in the embryo to specific regions of the neural plate. Expression in embryos was observed between prenatal days 9 and 14.5 throughout the neural plate at the level of the anterior head folds, and at the lateral tips of the neural plate in more posterior regions. After neural tube closure, *int*-1 expression was found to be restricted to the dorsal wall of the brain vesicles and spinal cord, the ventral wall of the midbrain and the diencephalon, and the lateral walls of the neuroepithelium at the midbrain-hindbrain junction. The *Drosophila* homologue of *int*-1 has been shown to be a segmentation polarity gene known as *wingless* (Rijsewijk et al., *Cell,* 50, 649—657). Zygotic lethal mutations of *wingless* show a replacement of the posterior part of each segment with an apparent mirror-image duplication of the anterior part. Viable alleles lead to a mirror-image duplication of a notum instead of a wing. Transcripts form *wingless* have also been found in the developing nervous system, but the main site of expression is in the epidermal parts of segments. Genetic evidence indicates abnormalities in pattern formation of segments only. Aberrant formation of the nervous system in *wingless* mutant has not been reported, suggesting that the mouse *int*-1 and *Drosophila wingless* gene may have different functional roles. Experiments in mouse will undoubtedly center on determining the effects of overexpression of *int*-1 on mouse development and on approaches to site-directed mutagenesis. *Terry Magnuson*

Deregulated c-*fos* Expression Interferes with Normal Bone Development in Transgenic Mice
U. Rüther, C. Garber, D. Komitowski, R. Müller, and E. F. Wagner
Nature (London), 325, 412—416, 1987 7-28

The *fos* oncogene is present in two mouse sarcoma viruses which both induce osteosarcomas. Evidence suggests that this gene may also function nonpathologically in processes of growth control and differentiation, perhaps by the involvement of its nuclear gene product in signal transduction. In this paper experiments are presented in which deregulated expression of c-*fos* was induced in transgenic mice.

Transgenic mice were produced which carried c-*fos* under the control of the glucocorticoid- and heavy metal-inducible metallothionein promoter. Animals carrying this construction showed little or no c-*fos* expression in ten organs examined, even when induced with cadmium. However, cycloheximide treatment or alteration of 3′ nontranslated sequences did permit stable c-*fos* expression in pancreas, kidney, brain, heart, muscle, lung, and salivary glands. Such overexpression of c-*fos* apparently resulted in animals with visible swelling on the tibia of the hind legs (see Figure 7-28). Histological investigations of these lesions showed the presence of normal processes of bone morphogenesis, but disordered remodeling processes, with bone formation far exceeding resorption. The lesions were neither malignant nor preneoplastic. Northern blot analysis suggested that exogenous c-*fos* expression occurred specifically in bone-forming cells and not in total bone marrow cells. The overexpression of c-*fos* in nonbone tissues was apparently without effect.

The authors believe that the formation of bone lesions is a direct result of deregulated c-*fos* expression. Therefore, they succeeded in altering a specific developmental process by perturbing the expression of a protooncogene.

♦It is often suggested that proto-oncogenes may play an important role in the genetic control of cellular differentiation and development. Many such genes are known to be expressed during gametogenesis and embryogenesis, but the functional significance of this expression is not yet understood. This paper presents data that are concerned with the role or the c-*fos* proto-oncogene in development and growth control. High levels of the c-*fos* gene

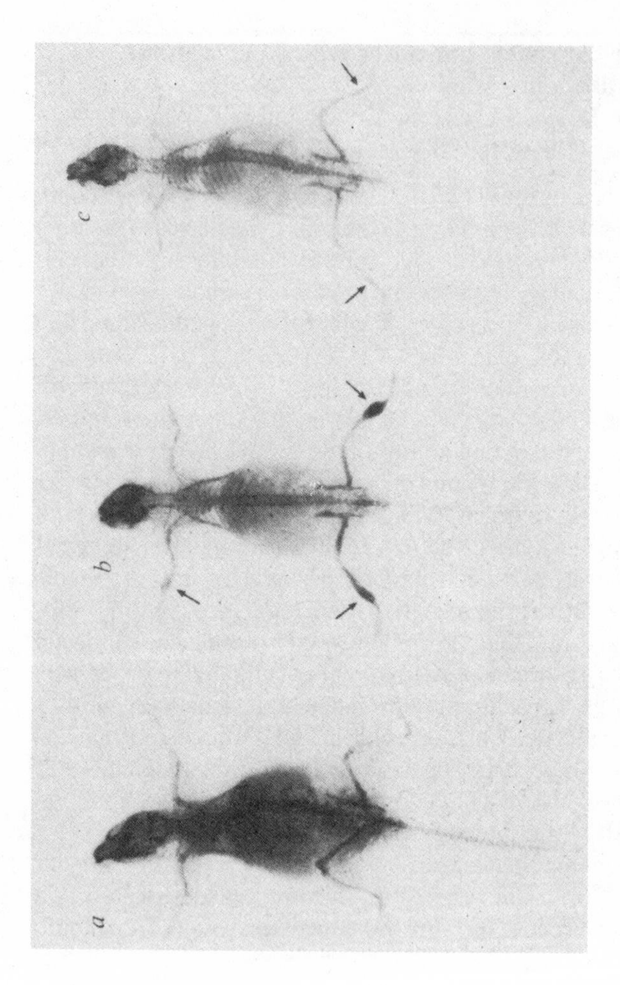

FIGURE 7-28. X-ray radiograph of normal and *c-fos*-expressing mice. *a*, Control C3H mouse; *b*, F2 offspring 209-4-2-41 showing clearly visible swellings; *c*, F2 offspring 209-4-2-44 without visible lesions. Arrows, bone abnormalities of different degrees. All three C3H mice were 15 weeks old. Methods: Radiographs of anesthetized mice were taken with a Siemens X-ray apparatus (B; 125/12/50R). The target-to-film distance was 105 cm and settings were 35 kV, 8 mA with 1.0 s exposure on Dupont, Cronex URF31 X-ray film. (Reprinted by permission from *Nature,* 325, 413. Copyright © 1987 Macmillan Magazines Ltd.)

product are normally detected in fetal membranes, yolk sac, and amnion, and as gestation proceeds, expression is found in fetal liver, adult bone marrow, and terminally differentiated blood granulocytes. These investigators have generated transgenic mice that express and exogenous c-*fos* construct in a deregulated manner. The construct introduced into the pronuclei of fertilized eggs contained the murine c-*fos* gene placed under the control of the glucocorticoid- and heavy-metal-inducible human metallothionein promoter. Analysis of mice carrying the construct showed that c-*fos* was expressed but that accumulation of mRNA appeared to be prevented by post-transcriptional regulatory mechanisms. Two independent mouse strains expressing exogenous c-*fos* in several tissues were produced, and when examined, approximately 40% of the males were small with visible swellings on the tibiae of both hind legs. Only 20% of the females showed similar abnormalities. The affected animals were healthy except for minor impairment of leg movements. Normal processes of bone morphogenesis were found to occur in these animals but remodeling progressed in an unordered manner with more bone formation occurring than bone resorption. The lesions were not malignant but consisted primarily of differentiated bone-synthesizing cells, and they seemed to be a direct consequence of deregulated c-*fos* expression. This paper represents a first step towards understanding the direct consequences of the c-*fos* proto-oncogene. *Terry Magnuson*

Homeobox Genes 8

INTRODUCTION

Perhaps the most exciting recent advance in the field of developmental biology involves identifying genes that direct the development of body segments. A unique feature of these genes is a highly conserved sequence of DNA known as the "homeobox".

First discovered in *Drosophila*, this DNA sequence has been shown to exist in a large number of genes from other species, including nonsegmented animals. The conserved and widely expressed nature of the homeobox suggests that the genes that contain these sequences may be fundamentally important in development; that their molecular homology might reflect functional homology.

The articles in this chapter all deal with homeobox-containing genes, their developmental expression, localization, and the nature of the proteins that they encode. They represent an area of intense current scientific energy.

An *enqrailed* Class Homeobox Gene in Sea Urchins
G. J. Dolecki and T. Humphreys
Gene, 64, 21—31, 1988 8-1

Homeoboxes are 180-base pair sequences which encode proteins believed to bind DNA and regulate gene expression. Two classes of homeoboxes exist in vertebrates, sea urchin, and insects: the *Antp* and *en* types (found in *Drosophila Antennapedia* and *engrailed* genes, respectively). The *Antp* genes of sea urchins have already been described by this group; the present paper describes the single *en* gene of sea urchins.

A *Tripneustes gratilla* genomic library was probed with a mouse *en* gene probe. Only a single clone hybridized under low stringency conditions. It was sequenced and found to be 62 to 73% identical to the nucleotide sequences of other *en* class homeoboxes. Its predicted amino acid sequence was 68 to 80% identical to that of other *en* class homeodomains. It had no intron, like that of mouse, and unlike that of *Drosophila en* boxes. When DNA from the sea urchin species *Sphaerechinus granularis* was reacted with the *T. gratilla en* probe, only a single band cross-hybridized, suggesting both that this sequence is conserved in sea urchins and that *S. granularis* too has but a single copy. Northern blot analysis of various adult *T. gratilla* tissues showed strong EU-HB-en transcription in Aristotle's lantern and weak transcription in ovary, testis, and coelomocyte. Northern analysis of *T. gratilla* embryos showed no sea urchin homeobox *en* (SU-HB-en) expression.

The authors speculate that the lack of embryonic expression of *en* genes in sea urchins, in contrast to strong embryonic expression in other organisms, may be because *en* transcripts are produced by ovaries for storage in the eggs and subsequent translation during development. They also note that the strong expression in Aristotle's lantern may be due to the presence of the nerve ring there, since *en* expression has been found in neural tissue of both flies and mice. Finally, Dolecki and Humphreys propose that the single *en* gene of sea urchins is the primitive evolutionary condition, and that two independent gene-duplication events led to the pairs of *en* class homeoboxes present in *Drosophila* and mice.

♦In this article the authors continue their studies on genes containing homeoboxes in sea urchins. They have previously characterized a *Antp*-like homeobox gene which is expressed during embryogenesis and is restricted to a subset of aboral ectoderm cells at the pluteus stage (Dolecki et al., *EMBO J.*, 5, 925—930, 1986). The function of the sea urchin *Antp*-like homeobox gene in these cells is not known. The present study concerns a second homeobox containing genes belonging to the *en* class. The major findings are that there is only a single gene belonging to the *en* class in the *T. gratilla* genome, that the gene is more closely related to the two mouse *en* class homeoboxes than to the two *Drosophila en* class homeoboxes, and that transcripts from this gene are seen only at appreciable levels in Aristotle's latern tissue, an adult tissue lining a calcified structure under the mouth. The significance of the transcript accumulation in this tissue is obscure though the authors suggest that it may be due to the presence of a nerve ring. Because the homeobox-containing genes are of obvious importance in a variety of developmental processes in *Drosophila* and other animals, an investigation of their function in sea urchins should prove valuable in gaining knowledge about cell determination and differentiation in this organism. By isolating and characterizing the genes and mRNAs for the two classes of homeoboxes, the authors have taken the first step towards understanding the function of these genes in sea urchins. *William H. Klein*

---

mec-3, a Homeobox-Containing Gene That Specifies Differentiation of the Touch Receptor Neurons in C. elegans

J. C. Way and M. Chalfie
Cell, 54, 5—16, 1988 8-2

In *Caenorhabditis elegans,* the set of six touch receptor neurons have been well characterized. Mutations that affect their development fall into three groups: their production, their specification, and their particular components. Only 1 gene, *mec-3,* is responsible for their specification. Mutants in this gene have normal cell production, but no differentiation of touch cells is seen. In this paper, the cellular phenotype of the mutants is further described, and studies of and with the cloned gene are reported.

Previous work had shown that in *mec-3* mutants, no cells exhibit any touch cell-specific features. Here, in loss-of-function mutants, the ALM touch cell is reported to assume novel features, while the animal is still completely touch insensitive. This implies that *mec-3* normally functions to repress this possible fate of the ALM cell. Further experiments began with the cloning of *mec-3* by transposon tagging. When transformed into germ cells, the clone isolated was able to restore wild-type phenotype to *mec-3* mutants. Transformants carrying extra copies of *mec-3* exhibit extra cells with *mec-3* dependent degeneration. Sequencing revealed that *mec-3* contains a homeobox and a region encoding acidic amino acids.

Taken together, these data suggest that *mec-3* controls the differentiation of the touch cells. The obvious implication is that the homeobox sequences function to bind DNA and the acidic regions (like in GAL4 and GCN4 of yeast) stimulate transcription. Thus, it is proposed that *mec-3* stimulates the transcription in touch cells of genes necessary for touch cell function.

♦This paper represents one of the first examples of a homeobox-containing gene, (*mec-3* in *C. elegans*) that is used exclusively in the nervous system (i.e., for something other than segment formation or segment identity in *Drosophila*). *Mec-3* appears to be a "selector" gene (to use the terminology of Garcia-Bellido in Cell Patterning, Ciba Foundation Symp., 29, 161—182, 1975) that activates "realization" genes whose products are the specific functional components of a cell. In the case of *mec-3* mutants, the affected cells are the set of six mechanosensory neurons that have been extensively characterized by Chalfie and colleagues (see Chalfie and Thomson, *J. Cell Biol.,* 93, 15—23, 1982; and Chalfie and Sulston, *Dev. Biol.,* 82, 358—370, 1981). Another homeobox gene that is required, like *mec*-3, only for proper neuronal differentiation is the *cut* gene of *Drosophila* (Blochlinger et al., *Nature (London),* 333, 629—634, 1988). *Joseph G. Culotti*

Temporal and Spatial Relationships Between Segmentation and Homeotic Gene Expression in *Drosophila* Embryos: Distributions of the *fushi tarazu, engrailed, Sex combs reduced, Antennapedia,* and *Ultrabithorax* Proteins

S. B. Carroll, S. DiNardo, P. H. O'Farrell, R. A. H. White, and M. P. Scott

Genes Dev., 2, 350—360, 1988 8-3

The development of segments in *Drosophila* embryos depends on the activity of genes of four classes. Gap, pair-rule, segment polarity, and homeotic genes are regulated in their expression by a network of interactions among genes of the different classes. In this article, the temporal and spatial expression of representative genes from each class was examined to define more precisely than previously the dynamics of accumulation of each gene product and the spatial boundaries between genes expressed simultaneously (see Figure 8-3).

This analysis involved *fushi tarazu (ftz)*, a pair-rule segmentation gene, *engrailed (en)*, a segment polarity gene, and *Sex combs reduced (Scr)*, *Antennapedia (Antp)*, and *Ultrabithorax (ubx)*, three homeotic genes. Their patterns of expression in embryos of different stages were determined with single-cell resolution, using double- and triple-label immunofluorescence methods. In some cases, sharp boundaries were visible. Sometimes these corresponded to parasegment boundaries, and sometimes they did not. Sharp boundaries sometimes blurred over time, while sometimes boundaries sharpened as development proceeded. The order of expression of the genes of various classes confirmed previous reports.

Coincident gene expression patterns were sometimes observed. These may imply regulatory relationships as may these data on the temporal order of expression of various genes in different cells. Because some cells contain one, and others several, homeotic gene products, the authors propose that combinatorial action of these proteins determines the differentiation of certain cells.

♦Development of the *Drosophila* embryo depends upon the sequential expression of at least four classes of zygotic genes. These genes are expressed only in restricted areas of the embryo and during a defined time window of embryonic development. The four classes of genes are, in approximate temporal order of expression, gap genes, which affect broad areas of the embryo; pair-rule genes, which affect development of alternate segments; segment polarity genes, which affect development of every segment; and homeotic genes, which affect the identity of cells originating within each segment, rather than the formation of the segments per se. For this study, double- and triple-label immunofluorescent staining of whole-mount embryos was used to analyze the expression of developmentally important genes. With this approach the authors determined the time of appearance of the different gene products and more importantly, the precise spatial relationship of expression with single-cell resolution. The following genes were analyzed: *fushi tarazu* (*ftz*) (pair-rule); *engrailed* (*en*) (segment polarity); and *Sex combs reduced, Antennapedia,* and *Ultrabithorax* (homeotics). All of these genes code for nuclear proteins containing a 60-amino-acid DNA binding sequence known as the homeodomain.

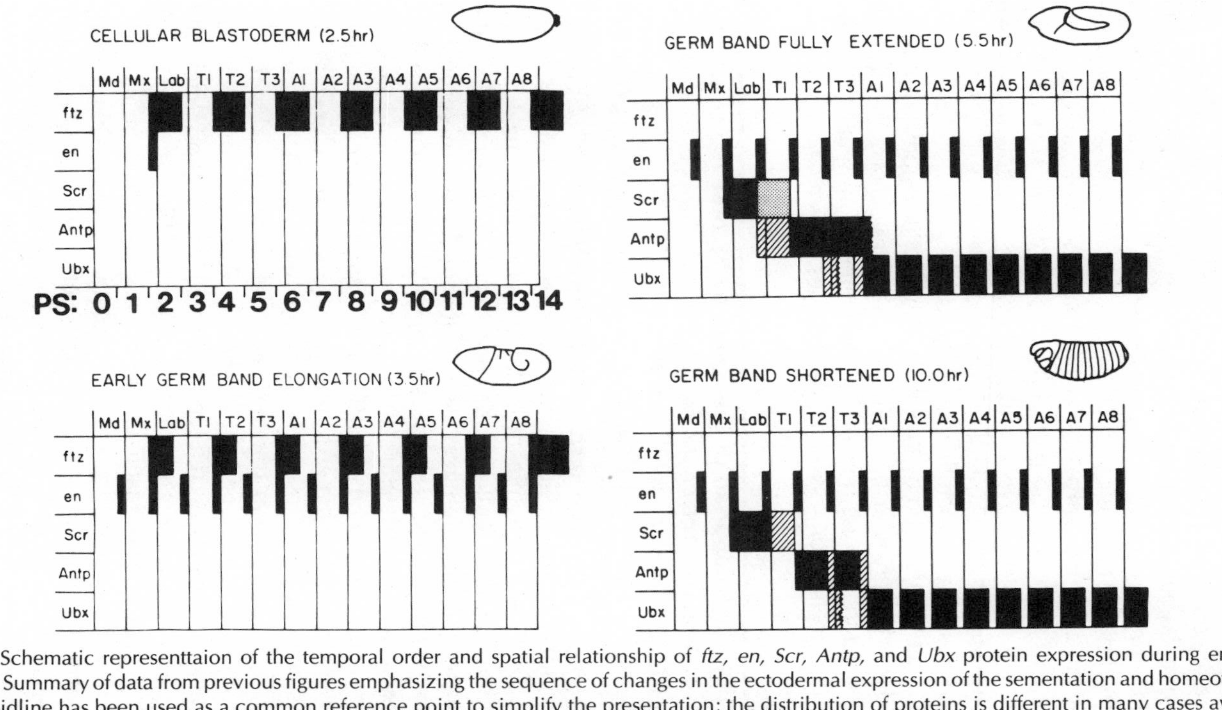

FIGURE 8-3. Schematic representtaion of the temporal order and spatial relationship of *ftz, en, Scr, Antp,* and *Ubx* protein expression during embryonic development. Summary of data from previous figures emphasizing the sequence of changes in the ectodermal expression of the sementation and homeotic genes. The ventral midline has been used as a common reference point to simplify the presentation; the distribution of proteins is different in many cases away from the ventral midline. The segmental position of high levels of protein expression is indicated by solid black shading and lower levels by hatching. The time and stage of development are indicated at the top of each panel. In some places, homeotic gene expression does not exhibit sharp discontinuities corresponding to *en*–expression boundaries (e.g., in PS 5, *Ubx* is not always coincident with *en* expression). These patterns are represented by wiggly lines. The homeotic protein patterns diagramed are correct for the epidermis only. The mesoderm and neural protein patterns may be quite different. (From *Genes Dev.*, 2, 350, 1988. With permission.)

In several instances the border of expression for a particular gene coincides cell for cell with the boundary of expression of an earlier expressed gene, suggesting a close interaction between pairs of genes. Frequently, the domain of expression of later genes respects a defined line of parasegmental borders that were probably established by the pair rule genes (such as *ftz*) and perhaps maintained by the segment polarity genes (such as *en*). In some cases the circular bands of homeotic gene expression vary along the dorsal-ventral axis, implying that genes that control dorsoventral differentiation are also part of this regulatory hierarchy. In some cases, expression of two genes was mutually exclusive, implying negative interactions, while in other cases different sets of genes are expressed in the same cell, suggesting cooperative interactions.

The information provided by this work, although descriptive in nature, should prove to be very valuable for the longer range goal of defining at the molecular level the precise network of gene interactions that lead to the differentiation of embryonic structures and tissues. *Marcelo Jacobs-Lorena*

Redesigning the Body Plan of *Drosophila* by Ectopic Expression of the Homoeotic Gene *Antennapedia*

A. Schneuwly, R. Klemenz, and W. J. Gehring
Nature (London), 325, 816—818, 1987 8-4

Antennapedia (Antp) deletion mutants of *Drosophila* can affect all three thoracic segments, transforming them toward phenotypes of more anterior segments. One class of dominant mutant shows transformation in the opposite direction, with adult antennae becoming second legs; this mutation seems to represent a dominant gain-of-function. In part, this prompted the hypothesis which is tested here: that ectopic overexpression of the *Antp* protein might cause the transformation of antennae into second legs.

A construction was created which fused the *Antp* protein-coding regions to the heat-inducible promoter of the heat shock gene *hsp70*. This DNA was introduced into wild-type flies by P-element mediated germ-line transformation. Several independent transformant lines were created which involved insertions on different chromosomes. After a heat shock, immunofluorescence demonstrated the presence of the *Antp* protein in the nuclei of cells throughout both embryos and larvae. When the heat shock was administered to third instar larvae, more than 50% of the flies had antennae transformed into apparently mesothoracic legs (see Figure 8-4). The transformations were always incomplete, but corresponded exactly to those of certain *Antp* mutants. This transformation only occurred when the heat shock occurred during the third larval instar.

These experiments suggest that *Antp* controls the development of the second thoracic segment. The data further imply that this gene specifies both dorsal and ventral parts of the mesothorax.

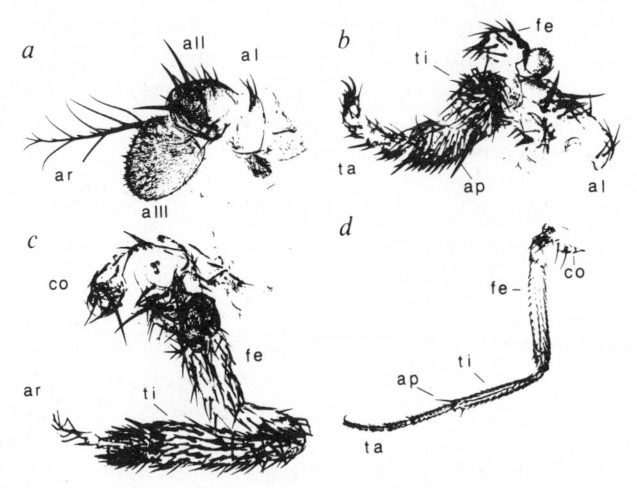

FIGURE 8-4. Induction of *Antp* phenotype by heat shock. *a,* Normal antenna (× 68). *b, c,* Two examples of antennae from heat-shocked flies. Heat shocks were for 2 h at 37° C on the fourth and fifth days of development (third-instar larvae). The main parts of the antennae are transformed into leg structures. In *b,* the arista is completely transformed into tarsal structures including claws, whereas in *c* more proximal structures are transformed. The presence of the apical bristle indicates the mesothoracic identity of the leg (×22) (×68). *d,* Normal mesothoracic leg (×22). Abbreviations: al, antennal segment I; all, antennal segment II; alll, antennal segment III; ar, arista; co, coxa; fe, femur; ti, tibia; ap, apical bristle; ta, tarsus. (Reprinted by permission from *Nature,* 325, 817. Copyright © 1988 Macmillan Magazines Ltd.)

Head and Thoracic Transformations Caused by Ectopic Expression of *Antennapedia* During *Drosophila* Development

G. Gibson and W. J. Gehring

Development, 102, 657—675, 1988 8-5

Previous work from this laboratory demonstrated that when the *Antennapedia (Antp)* gene is abnormally induced in all cells of *Drosophila* third instar larvae, antennae can be transformed into legs. This article extends that study, detailing the phenotypes of the transformations and the effects of *Antp* expression during various stages of development.

Flies were transformed using a construction consisting of *Antp* cDNA under the control of a heat shock promoter. When a heat shock induced *Antp* in all cells of embryos before germ-band retraction, a variety of abnormalities occurred. These included failure of head involution, appearance of novel cuticular structures in the head, and reductions in prothoracic denticle belts. When induced later in development, ectopic *Antp* expression sometimes caused lethality without apparent effects on the cuticle. During second and third instar larval stages, overexpression of *Antp* caused transformation of antenna toward mesothoracic leg. Such transformation required accumulation of high concentrations of *Antp* protein, suggesting that normal development can resist modest perturbations of homeotic gene expression. In addition, the most distal antennal structure was observed to be transformed by the earliest *Antp* inductions, and more proximal structures transformed by later inductions. During pupal stages, ectopic *Antp* expression was often lethal. Control experiments suggested that none of these effects was due to heat shock per se.

These experiments suggest that determination of the antennal disk proceeds in a distal-to-proximal direction and can be altered by strong, inappropriate *Antp* expression. The authors suggest that since small alterations in homeotic gene expression are apparently not harmful, they might serve to increase evolutionary flexibility.

♦Segmental identity in *Drosophila* is controlled by the activities of the homeotic genes. One such gene is *Antennapedia* (*Antp*), which is required for proper development of the thoracic segments. Loss-of-function mutants, which die as late embryos or early larvae, have thoracic segments transformed toward more anterior structures. Consistent with the interpretation that *Antp* is required for proper thoracic development is the finding that its mRNA and protein accumulate mainly in the prospective thoracic cells. Gain-of-function dominant mutants exist that show transformations in the opposite, anterior-to-posterior direction, namely from antenna into second leg. These mutations are believed to be due to fusion of the *Antp* coding region to unknown promoters, causing the ectopic expression of the wild type protein in the head. Schneuwly et al. linked the *Antp* coding region to the strong, inducible, and ubiquitously acting heat shock promoter. The main finding was that when the gene is heat induced during the third larval period, the emerging adults show transformations of parts of the antenna into second leg structures and transformation of regions of the hindhead into dorsal mesothoracic structures. These experiments show unambiguously that ectopic expression of a single gene at the appropriate developmental time is sufficient to cause transformation of head into more posterior thoracic structures. The paper by Gibson and Gehring describes in detail the phenotypes of the transformations. *Antp* induction during embryogenesis causes a series of abnormalities in head development, culminating in failure of the head to involute at the end of embryogenesis. Another finding was that depending on the time of *Antp* induction, a different region of the antenna transformed into leg; induction during early third instar causes transformation of the most distal antennal structures while induction at later times causes transformation of more proximal structures into the corresponding leg segments. Also of interest was the finding that for the antenna-to-leg transformation to occur, a relatively large output of the *Antp* protein is necessary at the ectopic location; only the strongest expressing transformed fly line shows clear-cut transformations and then only after multiple heat inductions of the gene. Finally, although the protein is ectopically expressed throughout the body of the transformant flies, only the head regions are affected.

This work shows how closely different developmental programs can be related, even when they lead to the formation of two body parts as different as head and thorax. Transformation of one structure into another can be accomplished by ectopic expression of a single homeotic gene. The results provide important clues to further our understanding of the principles underlying the specification of the *Drosophila* body plan. *Marcelo Jacobs-Lorena*

Divergent Homeo Box Proteins Recognize Similar DNA Sequences in *Drosophila*
T. Hoey and M. Levine
Nature (London), 332, 858—861, 1988 8-6

The *Drosophila* homeobox-containing gene, *even-skipped (eve)*, is one of the genes involved in the development of embryonic segmentation. It appears to exert its effect on morphogenesis by regulating its own expression and that of another segmentation gene, *engrailed (en)*. This regulation was explored in this series of experiments.

The binding of the *eve* protein to both the *eve* and *en* genes was examined, using a DNaseI footprint assay. Three regions of protection were observed in the 5′ region of the *en* gene and each contained at least one copy of a 10-bp consensus sequence, TCAATTAAAT. In the 5′ region of the *eve* gene, two binding sites were detected between -295 and -44 bp. These sequences were GC rich and did not resemble the binding sites in the *en* gene, despite comparable affinities for this protein. DNaseI protection assays were also used to compare binding affinities for these DNA sites of four homeobox-containing proteins: *eve, en, zerknult (zen),* and *paired (prd)*. These proteins all bound to the different sites with different relative affinities.

This work has demonstrated binding sites for the *eve* protein in the 5′ region of both the *eve* and *en* genes. The binding of these genes is consistent with *eve* regulation occurring at the transcriptional level.

♦The establishment of the *Drosophila* body plan depends critically on a small set of genes, whose hierarchical position in the developmental program is beginning to be understood in some detail. Many of these genes code for proteins containing a homeodomain, an amino acid sequence believed to confer DNA binding properties to the protein. Evidence exists suggesting that one of these genes, *even-skipped* (*eve*), regulates the activity of both *engrailed* (*en*) and itself. It is possible that regulation is at the transcriptional level and that the *eve* protein participates directly in this regulation by binding to the transcriptional regulatory sequences of the two genes. This hypothesis is addressed by the experiments described in this paper.

Full length proteins *eve, en,* and two other homeodomain-containing proteins, *zerknult* (*zen*) and *paired* (*prd*), were produced in bacteria from cDNA clones. These proteins were used in a binding assay to determine whether specific DNA sequences around the promoter region of the *en* and *eve* genes are recognized by the proteins. Specific binding was detected to one cluster of *en* DNA sequences and two clusters of *eve* DNA sequences 5′ of the transcription initiation sites. One protein (*eve*) bound specifically to sites on both *en* and *eve* genes. Interestingly, sequences recognized by the *eve* protein on the *en* and on the *eve* genes bear no relationship to each other (see "combinatorial" model below). Conversely, different proteins bound to the same sites but with relative affinities that varied over a 100-fold range (see "competition" model below). Thus, it appears that homeobox proteins recognize a broad range of DNA sequences, and that each protein binds with different affinity to each of these sites. This is consistent with at least two modes of transcriptional regulation by these genes: (1) "combinatorial", which requires the binding of more than one homeoprotein to the gene's regulatory region in order to promote activation or repression of transcription and (2) "competition", which assumes that two different proteins may compete for the same site to either promote or repress transcription and that the relative abundance and affinity of the two proteins determines the transcriptional activity of the gene being regulated. The present experiments and the available data in the literature are compatible with both of these models.

In the past few years a relatively large number of studies have described the hierarchical relationship of genes that determine the body plan of *Drosophila*. The present study represents one of the first to address possible mechanisms of cross regulation. *Marcelo Jacobs-Lorena*

A Homeobox-Containing Marker of Posterior Neural Differentiation Shows the Importance of Predetermination in Neural Induction

C. R. Sharpe, A. Fritz, E. M. DeRobertis, and J. B. Gurdon
Cell, 50, 749—758, 1987 8-7

Primary induction takes place during the formation of the amphibian axis and leads to formation of the neural tube from ectoderm that has been in contact with mesoderm. This paper describes a new, early molecular marker of neural induction.

A *Xenopus* gastrula cDNA library was screened with homeobox probes. One of the cDNAs obtained in this screen, XIHbox6, was used to probe a developmental polyA+ RNA Northern blot. Transcripts were detected by the early neurula stage and were maximal in the late neurula to tail bud stages. Embryos were dissected and probed to determine the region of expression. This gene was expressed only in the posterior two thirds of the nerve cord. Various regions of different stage embryos were cultured and then assayed for expression of this transcript. The gene was expressed only when both ectoderm and mesoderm were present in the culture, as would be expected if transcription of this gene were activated by induction. Stage 11 midgastrula dorsal mesoderm combined with stage 10 1/4 early gastrula ectoderm induced expression of this gene by the time control embryos reached stage 20—25. When the region of ectoderm used was varied (dorsal vs. ventral), it was observed that more dorsal ectoderm gave a stronger response.

This work confirms that neural development is dependent on the inducing effect of mesoderm on ectoderm. However, these results also indicate that the ectodermal portion that is destined to form neural tissue has a predisposition to respond to this induction.

♦The most important conclusion in this paper is that dorsal ectoderm is more easily induced than ventral ectoderm to differentiate into central nervous system when brought into contact with chordamesoderm. The basic experiment is to combine different sectors of competent ectoderm from stage 10.25 gastrula with dorsal mesoderm removed from stage 11 gastrula in a "sandwich", which is cultured and then assayed for the presence of mRNAs corresponding to a neural specific homeobox and to NCAM, which is neural specific in *Xenopus* at early stages. The enhanced sensitivity of dorsal vs. ventral ectoderm shows that to some extent the dorsal character of ectoderm is acquired prior to gastrulation. One strange result is that the dorsal half of the animal pole does not induce any better than ventral ectoderm, while the dorsal two-thirds does. There is clearly more work to be done, but some of the traditional notions about neural induction have already been called into question. *Thomas D. Sargent.*

Microinjection of Synthetic Xhox-1A Homeobox mRNA Disrupts Somite Formation in Developing Xenopus Embryos

R. P. Harvey and D. A. Melton
Cell, 53, 687—697, 1988 8-8

Microinjection of SP6-generated mRNAs into *Xenopus* embryos was used to investigate the function of the *Xenopus* homeobox-containing protein, Xhox-1A.

Four nanograms of Xhox-1A, a deleted form of this mRNA or beta-globin, was microinjected into *Xenopus* embryos at the time of the formation of the first cleavage furrow. The amount of the injected RNA that was retained at the neurula stage was 20-fold higher than the endogenous level of Xhox-1A. The level of deformed embryos injected with control RNAs was 13%, and with Xhox-1A was 51%. The Xhox-1A embryos were kinked in appearance. Histologically, there was disorganization and loss of the metameric patterning of the somite muscle tissue in embryos injected with excess Xhox-1A mRNA.

This paper presents evidence for the role of a homeobox gene, Xhox-1A, in vertebrate development. The results are consistent with involvement of this homeobox-containing gene in the specification of the segmental array of the somites. Therefore, segmentation in both vertebrates and flies may be controlled by homeobox products.

♦In spite of all the effort that has been expended on behalf of homeobox-containing genes in vertebrates, this paper contains the only experimental evidence that directly supports the belief that these genes have an important function in controlling vertebrate development. These authors injected synthetic mRNA corresponding to a *Xenopus* homeobox gene, Xhox-1A, into one cell of a two-cell stage embryo, and obtained reproducible disruption of somitogenesis on the injected side of the embryo. "Phenotypes" obtained by injecting substances into fertilized frog eggs must be very carefully evaluated, but the controls in this paper seem adequate. This appears to be successful application of a technique complementary to antisense gene ablation; forced ectopic expression of a tissue-specific gene by injecting synthetic mRNA. *Thomas D. Sargent*

Hox-1.6: a Mouse Homeo-Box-Containing Gene Member of the _Hox-1_ Complex

A. Baron, M. S. Featherstone, R. E. Hill, A. Hall, B. Gailliot, and D. Duboule

EMBO J., 6, 2977—2986, 1987 8-9

Many genes involved in the early development of *Drosophila* share a 180-bp DNA sequence, the homeobox. Similar sequences have been isolated from amphibians, rodents, and humans. In the mouse, 25 to 30 sequences exhibit high homology with the *Drosophila Antennapedia*-like *(Antp*-like) homeobox. *Hox-1.6* is a murine homeobox-containing member of the *Hox-1* complex which diverges more than other members of the complex from the *Antp*-like class. The authors have studied the transcription pattern of this gene.

Hox-1.6 is expressed during murine fetal development in an intestine-specific manner in adults, as well as in tumors or cell types showing early endoderm-like differentiation. Studies of embryonic partial *Hox-1.6* cDNA clones showed structural features similar to those of other *Drosophila* and vertebrate homeobox-containing genes, but *Hox-1.6* transcripts might have more complex splicing patterns. One cDNA clone contained a rather short open reading frame encoding a protein of about 14.5 kDa. Use of this clone as a probe for S1 nuclease mapping showed that different *Hox-1.6* transcripts were present in both embryonic total RNA and embryonal carcinoma cell cytoplasmic RNA.

These various transcripts may encode a set of related proteins. *Hox-1.6* provides the first example of multiple intronic sequences within a vertebrate homeobox-containing gene. Different proteins with the same DNA binding site — the homeodomain — but differing in composition of the amino-terminal region could have different functions during development or in adult tissues.

Expression of the Homeo Box-Containing Gene *En-2* Delineates a Specific Region of the Developing Mouse Brain
C. A. Davis, S. E. Nobel-Topham, J. Rossant, and A. J. Joyner
Genes Dev., 2, 361—371, 1988 8-10

Mammalian homeobox-containing genes may have a critical role in controlling development. The murine genes *En-1* and *En-2* contain a homeobox and flanking sequences homologous to those of the *Drosophila engrailed* and *invected* genes. If *En-2* has a role in pattern formation and differentiation, its expression should be spatially restricted in otherwise homogeneous tissues during development. The authors used *in situ* hybridization to determine whether the *En-2* transcript is expressed in a restricted area of the developing CNS from gestational day 8 to adulthood.

Transcripts were first detected in the neural folds of 8-d, 5-somite embryos. Expression continued into adulthood. Hybridization was observed only in the CNS and was limited to one band of the neural tube and to parts of structures that dveeloped from it, including the cerebellum, pons, periaqueductal gray, and colliculi. Hybridization was less extensive once cells had migrated from the germinal zone, and it was spatially more complex. In adults, expression of *En-2* was limited to specific groups of neurons.

The findings suggest that mouse *Hox* genes, including *En-2,* are important in development of the CNS through defining and maintaining spatial domains. This gene may be involved in the specification of different cell populations in the brain.

Specific Expression of the Hox 1.3 Homeo Box Gene in Murine Embryonic Structures Originating from or Induced by the Mesoderm
C. Dony and P. Gruss
EMBO J., 6, 2965—2975, 1987 8-11

Mammalian homeobox genes appear to closely resemble the respective *Drosophila* genes, but too little is known about the temporal and spatial expression of murine homeobox genes. Because little material is extractable from early-stage murine embryos, *in situ* hybridization studies are appropriate. The authors conducted an *in situ* analysis of the spatial and temporal expression profile of *Hox 1.3,* a member of a homeobox gene cluster on chromosome 6.

Hox 1.3 is expressed largely in mesoderm-derived or -induced embryonal structures. Expression is spatially limited to the thoracic region and to components of segmental origin such as the ribs and vertebrae and the precursor sclerotomes and somites. Expression also is observed in parts of the embryonal lung, gastric tissue, gut, and kidney. Expression is detected only while the thoracic structures are developing, on days 8 through 13, and not on day 18 when the embryo is mature.

Hox 1.3 is expressed in a distinct body part very early in murine development. Mesodermal structures expressing *Hox 1.3* are derived from somites or from region-specific mesoderm of the thoracic body part. A *Hox 1.3*-specific signal also is found in the CNS of the murine embryo. Formation of the murine spinal cord and brain is based on region-specific induction from the underlying mesoderm.

Expression of the Murine Homeo Box Gene Hox 1.5 during Embryogenesis

A. Fainsod, A. Awgulewitsch, and F. H. Ruddle

Dev. Biol., 124, 125—133, 1987 8-12

A majority of *Drosophila* homeobox genes exhibit specific spatial patterns of expression along the anteroposterior embryonic axis. The murine *Hox 3.1* gene is expressed in the spinal cord of newborn animals, and the *Hox 1.4* gene is expressed in the testes. The murine *Hox 1.5* gene contains a conserved homeobox, and it is expressed during embryogenesis and codes for multiple transcripts. The authors studied the pattern of expression of murine *Hox 1.5* by *in situ* hybridization to frozen sections of mouse embryos, and by Northern blot hybridization to RNA samples from embryos and adult spinal cord.

In situ hybridization of single-stranded RNA probes to mouse embryo sections showed a specific spatial distribution of *Hox 1.5* transcripts in 8.5 to 12.5-d embryos. Expression was limited to the central nervous system (CNS), with an anterior boundary in the hindbrain; it extended posteriorly through caudal areas of the spinal cord.

The murine *Hox 1.5* gene is expressed in a spatially restricted manner during embryonic development, in the CNS. Its expression resembles that observed in *Drosophila* homeotic genes. Studies correlating *Hox 1.5* expression with landmarks of neural morphogenesis may elucidate its role during this stage of development.

Coding Sequence and Expression of the Homeobox Gene *Hox 1.3*

M. Fibi, B. Zink, M. Kessel, A. M. Colberg-Poley, S. Labeit, H. Lehrach, and P. Gruss

Development, 102, 349—359, 1988 8-13

Hox 1.3 is a murine homeobox-containing gene that is a member of the *Hox 1* cluster on chromosome 6. The authors isolated a cDNA of the *Antp* homologous region *Hox 1.3,* and predicted the primary structure of the *Hox 1.3* protein by deduction from an open reading frame containing the homeobox.

The protein sequence contains 270 amino acids. The *Hox 1.3* protein exhibits extensive sequence homology with the murine homeodomain protein *Hox 2.1,* encoded on chromosome 11. In murine embryonic development, peak expression of *Hox 1.3* occurs in 12-d tissue. An abundant *Hox 1.3*-specific 1.9-kb RNA is found in F9 cells induced for parietal endoderm differentiation. Negative correlation between transformation and *Hox 1.3* expression is noted in 3T3 fibroblasts. Untransformed 3T3 cells carry abundant 1.9-kb *Hox 1.3* RNA, while methylcholanthrene-transformed cells express only low levels of this RNA.

There is a fundamental similarity between all class I homeodomain vertebrate proteins described to date. The murine pair *Hox 1.3* and *Hox 2.1* represents a high level of conservation, indicating that gene duplication has occurred at some point in evolution.

Homeobox Gene *Hox 1.5* Expression in Mouse Embryos: Earliest Detection by *in situ* Hybridization is During Gastrulation

S. J. Gaunt

Development, 101, 51—60, 1987 8-14

In mammalian homeobox genes functionally similar to those of *Drosophila,* restricted spatial expression early in development is expected, possibly during the time that cell determination occurs along the anteroposterior axis. Earlier studies of the murine homeobox gene *Hox-1.5* showed it to be expressed in posterior ectoderm and mesoderm in 7- to 8-d embryos, and in the nervous system at 9 1/2 d. Further stages of development now have been analyzed by *in situ* hybridization.

At 7 1/2 to 9 1/2 d of development, *Hox-1.5* expression was restricted to the posterior regions of the embryo. Transcription in presomitic and somitic mesoderm declined relative to that in the overlying neural ectoderm at this stage. The gene was expressed posterior to the hindbrain boundary in 10 1/2- to 12 1/2-d embryos, but not in newborn mice. Embryos as old as 7 1/4 d showed no evidence of *Hox-1.5* transcripts.

Hox-1.5 is expressed during the time that tissues are determined along the anteroposterior axis of the mouse. Initial spatial restriction of *Hox-1.5* expression might conceivably depend on a morphological gradient across the cells of the epiblast or newly formed mesoderm layer. The process may involve the establishment of permeable intercellular channels for transfer of metabolites.

Spatially Restricted Patterns of Expression of the Homeobox-Containing Gene *Hox 2.1* During Mouse Embryogenesis

P. W. H. Holland and B. L. M. Hogan
Development, 102, 159—174, 1988 8-15

Because the murine *Hox 2.1* gene contains a homeobox sequence, it may be involved in the control of embryonic patterning or positional specification. The authors used *in situ* hybridization to determine the pattern of *Hox 2.1* expression during murine embryogenesis.

At 8.5 d, *Hox 2.1* was expressed at a low level in the posterior neuroectoderm and mesoderm, as well as in the neuroectoderm of the presumptive hindbrain. The pattern at 12.5 d was one of restricted anteroposterior expression from the hindbrain through the length of the spinal cord, predominantly dorsally. At 13.5 d, expression was limited to the occipital and cervical regions. *Hox 2.1* RNA also was found in the embryonic lung, stomach, mesonephros, and metanephros. It was present in the myenteric plexus, dorsal root ganglia, and nodose ganglion, as well as in mature granulocytes.

Apart from a regulatory role in anteroposterior positional specification in the neuroectoderm and mesoderm, *Hox 2.1* may function during organogenesis.

En-1 and *En-2,* Two Mouse Genes with Sequence Homology to the *Drosophila engrailed* Gene: Expression During Embryogenesis

A. L. Joyner and G. R. Martin
Genes Dev., 1, 29—38, 1987 8-16

The murine genes *En-1* and *En-2* have sequence homology with the *engrailed* and *invected* genes of *Drosophila.* The homologous sequences may encode a segment of 107 amino acids, including a central 60-amino acid homeobox. Nearly three fourths of amino acids within the homologous regions of the genes are identical in the mouse and *Drosophila.* The authors attempted to define the tissue specificity of *En-1* and *En-2* during embryogenesis.

Transcripts from both genes were found in RNA samples from teratocarcinoma cells and from embryos at 9.5 to 17.5 d of development. Transcripts were abundant in RNA extracted from the posterior part of the fetal brain, but much less abundant in RNA from other fetal tissues, including the anterior part of the brain and the spinal cord. Recombinant inbred strain analyses showed that — unlike their *Drosophila* counterparts — *En-1* and *En-2* were unlinked. *En-1* is in the central part of murine chromosome 1, while *En-2* is in the proximal part of chromosome 5. Both genes mapped near mutations that cause developmental abnormalities.

These mouse genes may be homologues of the *Drosophila en* and *inv* genes with regard to biochemical function, but it is not clear whether they help control embryogenesis in mammals as they do in *Drosophila.* Any functional significance in the close linkage of the *Drosophila* genes is not conserved in the mamalian genes.

Developmental and Spatial Patterns of the Mouse Homeobox Gene, *Hox 2.1*

R. Krumlauf, P. W. H. Holland, J. H. McVey, and B. L. M. Hogan

Development, 99, 603—617, 1987 8-17

The *Hox 2.1* gene is part of a cluster of homeobox-containing genes on murine chromosome 11. Analysis of *Hox 2.1* cDNAs from a mouse embryo library predicted that the gene encodes a 269-amino acid protein which contains a homeobox 15 amino acids from the carboxy terminus. A second partially conserved region, present in other genes containing homeoboxes, is 12 amino acids upstream of the homeodomain.

RNA transcripts of *Hox 2.1* are found in fetal lung, spinal cord, kidney, gut, spleen, liver, and visceral yolk sac. It appears that *Hox 2.1* is regulated independently in various tissues. Transcripts from other loci have extensive homology with *Hox 2.1* in sequences outside the homeobox. *In situ* hybridization indicated that *Hox 2.1* transcripts are localized in the spinal cord in an anteroposterior gradient extending from the hindbrain. In the lung, mesodermal cells surrounding the branching epithelial-cell layer accumulated high levels of *Hox 2.1* transcripts.

Direct confirmation of a functional embryonic role of *Hox 2.1* and other vertebrate homeobox-containing genes remains to be obtained. This question will be addressed using transgenic mice and antibody probes to examine protein expression in the early embryo.

Pattern of Transcription of the Homeo Gene *Hox-3.1* in the Mouse Embryo

H. LeMouellic, H. Condamine, and P. Brûlet

Genes Dev., 2, 125—135, 1988 8-18

The authors cloned and sequenced a cDNA from the murine homeogene *Hox-3.1,* and studied patterns of transcription of the gene at days 8.5 to 14.5 of mouse embryogenesis. The cDNA from the *Hox-3.1* locus was isolated from a 10.5-d mouse embryo cDNA library. Spatial distribution of *Hox-3.1* gene transcripts was monitored from late gastrulation to embryonic day 14.5 by *in situ* hybridization.

When first detected in 8.5-d embryos, the transcripts were present in all tissues of the posterior end of the embryo. Subsequently, the distribution of transcripts was progressively restricted and tissue specific. By 12.5 d, transcripts were localized predominantly in the neural tube region above the heart. An area of neural tube posterior to the brain and mesenchymal cells about lung tubules were highly positive.

The early pattern of transcription of *Hox-3.1* is consistent with a regionalizing role for this gene. In the chicken, the somitomeres of the presomitic mesoderm already carry the information needed for future somite organization, including anteroposterior polarity.

Murine *Hox-1.7* Homeo-Box Gene: Cloning, Chromosomal Location, and Expression

M. R. Rubin, W. King, L. E. Toth, I. S. Sawczuk, M. S. Levine, P. D'Eustachio, and M. C. Nguyen-Huu
Mol. Cell. Biol., 7, 3836—3841, 1987 8-19

Genomic DNA from vertebrates contains several copies of homeoboxes. Expression of several of these homeoboxes is localized to specific regions of the mouse embryo, in analogy with homeobox gene expression in the *Drosophila* embryo. The authors identified a new murine homeobox, *Hox-1.7,* in a rare cDNA from F9 teratocarcinoma stem cells. It is 68 and 72% homologous with the *Drosophila Antennapedia (Antp)* and *iab-7* homeoboxes, respectively.

A major 2.5-kb transcript and several minor transcripts were detected by Northern blot analysis in adult tissues and in midgestational embryos. The posterior spinal cord was a prominent site of *Hox-1.7* expression in 12.5-d embryos. Transcripts were not found in embryonic brain tissue. Somatic cell hybrids mapped the *Hox-1.7* gene to chromosome 6. The distribution patterns of restriction fragment length polymorphisms in recombinant inbred mouse strains mapped *Hox-1* and *Hox-1.7* close to two mouse loci affecting morphogenesis, postaxial hemimelia *(px)* and hypodactyly *(Hd)*.

It remains to be learned when in embryogenesis the *Hox-1.7* gene is first expressed. Like many other homeobox genes, *Hox-1.7* is strongly expressed in embryonic and adult spinal cord, but in the embryonic cord it is preferentially expressed posteriorly. The physical relationship of *Hox-1.7* to the rest of the cluster is uncertain.

Isolation and Expression of New Mouse Homeobox Gene

P. T. Sharpe, J. R. Miller, E. P. Evans, M. D. Burtenshaw, and S.J. Gaunt
Development, 102, 397—407, 1988 8-20

A homeobox-containing clone isolated from an adult mouse kidney cDNA library was found on DNA sequence analysis to be a new isolate, *Hox-6.1.* A genomic clone containing *Hox-6.1* contains another putative homeobox sequence, *Hox-6.2,* within 7 kb of *Hox-6.1. In situ* hybridization of mouse metaphase chromosomes showed the *Hox-6* locus to be on chromosome 14. The predicted protein sequence of the homeobox of *Hox-6.1* was 100%

homologous with the *Xenopus Xebl* homeobox and the human *c8* homeobox.

RNase protection assays showed that *Hox-6.1* transcripts were present in embryos at 9 1/2- to 13 1/2-d gestation, and in extraembryonic tissues at day 9 1/2. In adults, expression was detected in the kidney and testis, but not in the liver, spleen, or brain. A major transcript in kidney was 2.7 kb long, while in embryos the major transcript was 1.9 kb. Expression of *Hox-6.1* was apparent in the spinal cord and prevertebral column in day 12 1/2 embryos, and in the posterior mesoderm and ectoderm at day 8 1/4. The anterior boundary of expression was just behind the hindbrain.

It would seem that *Hox-6.1,* like *Hox-1.5,* is part of a series of homeobox genes whose expression provides cues to positioning during determination of tissues along the body axis.

Region-Specific Expression of Mouse Homeobox Genes in the Embryonic Mesoderm and Central Nervous System

L. E. Toth, K. L. Slawin, J. E. Pintar, and M. C. Nguyen-Huu

Proc. Natl. Acad. Sci. U.S.A., 84, 6790—6794, 1987 8-21

Copies of homeoboxes are found in the mammalian genome, but it is not clear whether they are components of morphogenetic loci, as in *Drosophila.* The authors used *in situ* hybridization to define spatial patterns of expression of two mouse homeobox genes in the midgestational embryo. The genes, *Hox 1.2* and *Hox 1.4,* are 20 kb apart on mouse chromosome 6.

Hox 1.2 transcripts were found chiefly in the posterior myelencephalon, the cervical CNS, and several thoracic prevertebrae. *Hox 1.4* transcripts were present mainly in the posterior myelencephalon and the cervical CNS. Within the latter area, transcripts were abundant in the mantle layer than in the ependymal layer, and higher levels were present dorsally than ventrally.

The restricted pattern of expression of these genes along the rostrocaudal axis of the mouse embryo resembles the pattern of expression of *Drosophila* homeoboxes in the embryo and larvae. It is possible that functional similarities exist between *Drosophila* and mammalian homeobox genes, despite different developmental strategies. Specifically, mammalian homeobox genes may have a role in determining regions in the embryonic CNS.

Region-Specific Expression of Two Mouse Homeo Box Genes

M. F. Utset, A. Awgulewitsch, F. H. Ruddle, and W. McGinnis

Science, 235, 1379—1382, 1987 8-22

Several murine genes contain sequences homologous to the homeobox sequences of *Drosophila* homeotic and segmentation genes. The authors studied the distribution of transcripts of two murine homeobox genes, *Hox-*

2.1 and *Hox-3.1,* during the latter third of prenatal development.

Transcripts from *Hox-2.1* and *Hox-3.1* were localized along the rostrocaudal axis of the developing central nervous system (CNS). Patterns of expression were quite similar to those of *Drosophila* homeotic gene expression. Each gene had a unique anterior boundary of expression within the CNS, which persisted through prenatal development. *Hox-3.1* expression was limited to the spinal cord caudal to the third cervical vertebra. *Hox-2.1* expression was restricted to the medulla and spinal cord, with the highest levels in the medulla.

These observations are consistent with a patterning function for *Hox-2.1* and *Hox-3.1,* similar to that performed by the *Drosophila* genes. *Hox-2.1* expression may help determine the fate of cells in the area of the medulla and anterior spinal cord, while *Hox-3.1* plays a determinative role posterior to the third cervical vertebra.

Differential Expression of the Mouse Homeobox-Containing Gene *Hox-1.4* During Male Germ Cell Differentiation and Embryonic Development

D. J. Wolgemuth, C. M. Viviano, E. Gizang-Ginsberg, M. A. Frohman, A. L. Joyner, and G. R. Martin

Proc. Natl. Acad. Sci. U.S.A., 84, 5813—5817, 1987 8-23

When the 180-bp homeobox domain of the *Antp* gene of *Drosophila* was used to screen a mouse testicular cDNA library, the homeobox-containing gene *Hox-1.4* was isolated. This gene is expressed specifically and abundantly in the adult mouse testis. *Hox-1.4* is one of a cluster of at least six homeobox-containing genes mapping to mouse chromosome 6.

Hox-1.4 transcripts about 1.4 kb long were found in adult testis, but not in testes of 17- to 20-d embryos or in neonatal testes. Studies of RNA from testes of mutant mice deficient in germ cells confirmed that *Hox-1.4* expression in the testis is germ cell specific. Gene expression was localized to germ cells that had entered into and progressed beyond the meiotic prophase stage of differentiation. *Hox-1.4* transcripts larger than those in germ cells were found by examining RNA from teratocarcinoma-cell cultures and mouse embryos at 10.5- to 16.5-d gestation. In the midgestational embryo, *Hox-1.4* expression is most evident in the spinal cord.

The highly restricted tissue specificity of *Hox-1.4* expression in the adult mouse is unique among homeobox-containing genes studied to date. However, other *Hox* loci are expressed in the adult testis. Studies of functional differences among *Hox-1.4* transcripts will help elucidate the role of this gene in male germ cell and embryonic development.

♦The highly conserved nature of the homeobox sequence in *Drosophila* has led many investigators to look for similar sequences in mouse. To date, of the 25 to 30 possible *Antennapedia*-like (*Ant*-like) and *engrailed*-like (*en*-like) homeobox sequences detected, 19 have been cloned. Sixteen have been shown to reside at loci that code for transcripts expressed during embryogenesis. Two major clusters of the *Ant*-like homeoboxes have so far been found, the *Hox-1* complex (chromosome 6) which contains six, possibly seven homeoboxes, and the *Hox-2* complex (chromosome 11) which contains six homeoboxes. The remaining *Ant*-like homeoboxes (*Hox 3, 4,* and *6*) are on chromosomes 15, 12, and 14, respectively. There are two *en*-like homeoboxes located on chromosomes 1 and 5. To address whether homeobox-containing genes play a role in development, temporal and spatial expression patterns have been studied during the past year. In general, the onset of transcription occurs at embryonic day 7.5 to 8.5. At this time, transcripts are detected in the embryonic ectoderm and mesoderm. After day 8.5, the pattern of expression undergoes a change from a more diffuse localization over the posterior ectoderm and mesoderm to a more defined pattern of expression that extends anteriorly. The highest levels of expression usually occur during midgestation and in many cases is completed by day 18.5. Analysis of expression by *in situ* studies reveals distinct anterior boundaries with levels of expression declining in posterior regions of the embryo. Certain homeobox genes are expressed in the developing lung, mesonephros, and metanephros but only in the mesenchymal cells. The areas of expression in the adult usually correspond to tissues within which these homeobox genes were expressed during development. The one exception is the testis where several of these genes are expressed in the adult. The role of the homeobox genes in mouse development remains to be established. The overlapping domains of expression detected by *in situ* analyses suggest the possibility of anterior-posterior positional cues. Future work will undoubtedly concentrate on transgenic experiments to produce improper expression of these genes, and also on site-directed mutagenesis by homologous recombination. *Terry Magnuson*

Morphogenesis and Pattern Formation 9

INTRODUCTION

Now that all the pieces have been identified and explored, it's time to try and complete the puzzle — a task that is easier said than done.

The accurate and appropriate arrangement of individual cells within an entire organism requires critical control. How any individual cell knows where it is within an organism, what it should be, and what its neighbors should be are intriguing questions about which little is currently understood.

A role for chemical morphogens has long been suggested and recent work indicates that such molecules may in fact exist, and that their chemical natures have been determined.

In addition to small morphogenetic molecules, there is evidence that gene products, cell surface molecules, and perhaps even proto-oncogene products may play a role in determining polarity and ultimately influence the pattern formation in developing embryos.

The papers in this chapter examine molecules that may have morphogenetic roles and look at some interesting cellular features of pattern formation.

Genesis of a Spatial Pattern in the Cellular Slime Mold *Polysphondylium pallidum*
G. Byrne and E. C. Cox
Proc. Natl. Acad. Sci. U.S.A., 84, 4140—4144, 1987 9-1

The mature fruiting body of the cellular slime mold *Polysphondylium pallidum* is composed of a central stalk along which several sets of branches are arrayed at regular intervals. Formation of this pattern begins with the segregation of whorls from the base and the formation of tips around the whorl. Each tip organizes the formation of a branch. The spatial distribution of the tip-specific antigen, Pg101, was quantified by Fourier analysis.

A monoclonal antibody against Pg101 was used to stain whorls. A band of increased fluorescence was detected along the equatorial surface prior to tip formation. The band fragmented into clusters of spots, which will be the positions of tip formation. Fourier analysis was used to analyze the pattern on the equator of 15 whorls. During the initial stages, the band was determined to be randomly distributed. Over time this evolved into a periodic prepattern. Prior to tip formation, the whorls displayed a periodic staining pattern that sharpened over time.

Fourier transformation of Pg101 density profiles indicated that tip-specific antigen was initially expressed randomly and was transformed into a periodic prepattern. As this pattern was established prior to tip formation, a reaction-diffusion-like mechanism may be used to establish the prepattern.

♦The mechanism of side branch formation in the cellular slime mold *Polysphondylium pallidum* was investigated using monoclonal antibodies that recognize the same antigen that is expressed in association with the developing tips of the branches. The monoclonal antibody anti-Pg101 detects its cognate antigen Pg101 at high levels at the anterior tip of the developing fruiting body. Fourier analysis indicates that tip-specific antigens are initially expressed randomly around the whorl. The transformation of the antigen staining into a periodic prepattern occurs uniformly around the circumference of the whorl. This occurs long before any morphological differentiation can be observed. Thus, it is suggested that the antigen expression is not completely homogeneous. The initial small random differences are transformed into a periodic distribution by autocatalysis and lateral inhibition. The process of pattern formation was modeled using a series of reaction-diffusion equations. The modeling results in a gradual process of pattern formation that is similar to that observed while monitoring the expression of the tip-specific antigen Pg101. The methods described here may be useful in analyzing pattern formation in other organisms where symmetry breaking appears to be a mechanistic component.
Stephen Alexander

Chemical Structure of the Morphogen Differentiation-Inducing Factor from *Dictyostelium discoideum*

H. R. Morris, G. W. Taylor, M. S. Masento, K. A. Jermyn, and R. R. Kay
Nature (London), 328, 811—813, 1987 9-2

Morphogens are embryonic signaling molecules involved in establishing developmental patterns. The *Dictyostelium* morphogen DIF-1 induces stalk cell differentiation from isolated amoebas. Purified DIF-1 was used to investigate its structure.

The structure of DIF-1 was examined through microchemical derivitization followed by functional bioassay, UV absorbance, fast atom bombardment mass spectroscopy, proton NMR spectroscopy, and synthesis. Synthetic compounds were compared to natural DIF-1 by HPLC, GC, mass

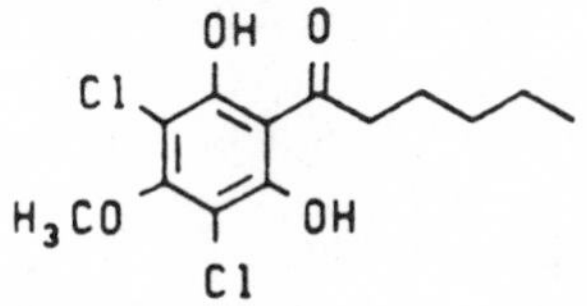

OH O
Cl
H_3CO OH
Cl

FIGURE 9-2. The structure of the morphogen DIF-1 was determined by microchemical analysis, spectroscopy, and synthesis to be 1-(3,5-dichloro-2,6-dihydroxy-4-methoxyphenyl)-hexanone. (Reprinted by permission from *Nature*, 328, 813. Copyright © 1987 Macmillan Magazines Ltd.)

spectroscopy, NMR, and bioassays. The structure of DIF-1 was defined to be 1-(3,5-dichloro-2,6-dihydroxy-4-methoxyphenyl)-1-hexanone (Figure 9-2).

DIF-1 is a new type of biologically active compound. Analysis of the biosynthesis and function of DIF-1 will be facilitated by knowledge of its structure and availability of synthetic DIF-1 and its analogues.

♦Isolated *Dictyostelium* amoebae can be induced to differentiate into stalk cells by the addition of a dialysable factor produced by cells undergoing normal development. The structure of this morphogen (DIF-1) has now been determined and the synthetic morphogen is available for biological studies. The elucidation of the structure opens the way to the analysis of the biosynthesis of DIF and its role in the establishment of pattern and cell-type specific gene expression in this organism. *Stephen Alexander*

Direct Induction of Dictyostelium Prestalk Gene Expression by DIF Provides Evidence That DIF is a Morphogen

J. G. Williams, A. Ceccarelli, S. McRobbie, H. Mahbubani, R. R. Kay, A. Early, M. Berks, and K. A. Jermyn
Cell, 49, 185—192, 1987 9-3

HM44 is a *Dictyostelium* mutant that is defective in the production of DIF and, therefore, produces no stalks. *In vitro* induction of HM44 cells with exogenous DIF was used to isolate a cDNA clone, pDd63, derived from an mRNA that is rapidly induced by DIF.

The pDd63 message can be detected within 15 min of DIF addition, and its continued synthesis is completely dependent on the presence of DIF. *In vitro* transcription assays with isolated nuclei indicated that DIF induction of pDd63 occurred at the level of transcription. A corresponding genomic clone was isolated and sequenced. The central portion of the gene was composed of a tandem array of a highly conserved, cysteine-rich 24 amino acid repeat. The extreme amino terminus of the protein-coding region contained a putative signal sequence. The kinetics of pDd63 message accumulation were consistent with induction by DIF during normal development. Prestalk cells were separated from prespore cells by microdissection and density gradient centrifugation. The pDd63 mRNA was highly enriched in prestalk cells.

These results suggest that DIF acts as a morphogen during normal *Dictyostelium* development and induces the expression of the pDd63 gene. This gene can now be used to identify *trans*-acting factors that mediate DIF induction of transcription during development. Using pDd63 as a marker of prestalk differentiation, prespore cells appeared to differentiate first. It is possible that prespore differentiation is necessary for prestalk differentiation to proceed.

♦The mechanism of action of DIF, a specific *Dictyostelium* morphogen, has been studied by showing that specific proteins are expressed when purified DIF is added to mutant cells (strain HM44) which are developmentally arrested due to their inability to express DIF. The gene for one of these proteins now has been cloned and its pattern of expression during normal development has been characterized. The gene is expressed at maximal levels during slug migration and the mRNA is enriched in the prespore cells. There does not appear to be expression in the prespore population at any time of development. The gene is expressed at the tipped aggregate stage several hours after prespore genes. Thus, prespore gene expression may be a prerequisite for prestalk gene expression. The specific induction of the gene, which is normally expressed exclusively in the prestalk zone, strongly supports the contention that DIF is a specific morphogen in *Dictyostelium*.
Stephen Alexander

Movement of the Multicellular Slug Stage of *Dictyostelium discoideum*: an Analytical Approach

E. J. Breen, P. H. Vardy, and K. L. Williams

Development, 101, 313—321, 1987 9-4

The amoebas in a migrating *Dictyostelium* slug move through a continuously produced extracellular slime trail. Migrating slugs were filmed to observe their interactions with the substratum.

In migrating slugs of the strain WS380B, the tip was raised, projected, and lowered approximately every 12 min in a coordinated fashion, such that the length of the slug remained constant. An indentation around the perimeter of the slug, called the collar, formed immediately after the tip was lowered to the substratum and became stationary as the tip was projected. When the slime trail was treated with FITC-labeled monoclonal antibody MUD50, which recognizes a glycoconjugate antigen, footprints were observed. These footprints corresponded to the appearance of the collar in moving slugs. The glycosylation defective mutation *modB* was transferred into the WS380B background to create strain HU2421, which lacks the epitope recognized by MUD50. HU2421 slugs migrated slowly and had a boundary or waist between the anterior and posterior regions. In the HU2421 strain, collars were not stationary, but moved backward relative to the substratum and halted at the waist.

In the squeeze-pull model of slug migration, it is proposed that some of the slug cells are specialized for locomotion and gain traction on the substratum. The collar represents a stationary aspect of slug migration, which corresponds to slime trail footprints, and lends support to this model. The reduced migration rate of *modB* mutants appears to be due to loss of traction. Therefore, the antigen recognized by MUD50 may be responsible for creating traction between amoebas and substratum.

♦Developing aggregates of *Dictyostelium* can, under the appropriate environmental conditions, transform into migrating slugs and move away from the site of aggregation toward sources of light or heat. The slug produces and moves through an extracellular sheath that is left behind as a collapsed "slime" trail. Analysis of these trails with monoclonal antibodies (MUD50 which recognizes a carbohydrate epitope on a family of glycoproteins) suggests that the slug interacts with the sheath to generate motive force. A collar-like structure forms behind the tip just after the tip is lowered. The collar is stationary at the substrate as the tip is projected upwards and forwards. Staining the sheath with the MUD50 antibody reveals a series of "footprints" which show where the slug has made contact with the sheath. The position of the MUD50 staining in the sheath matches with the appearance of the collar. A similar analysis was performed on a recombinant strain that had the *modB* mutation placed in the WS380B genetic background. Unlike the wild-type slugs, the collar of the *modB* strains slips backwards and ends up at a distinct waist region. Retarded migration appears to be the result of the lack of traction due to the collar slipping. The results are discussed in terms of models for slug migration and the biochemical nature of the glycoproteins recognized by the MUD50 antibody. *Stephen Alexander*

Selective Disruption of Gap Junctional Communication Interferes with a Patterning Process in Hydra

S. E. Fraser, C. R. Green, H. R. Bode, and N. B. Gilula
Science, 237, 49—55, 1987 9-5

In the hydra, the presence of a head inhibits head formation elsewhere in the body. This head inhibition is assumed to be due to the diffusion of a labile head inhibitor, produced by the head and degraded by other tissues. It is possible that gap junctional communication is involved in transmission of the head inhibitor throughout the hydra body.

Antibodies to the major rat liver gap junction protein were used for indirect immunofluorescence on hydra cells. The distribution of antigen was as expected for a hydra gap junction protein. When these antibodies were microinjected into hydra epithelial cells, they blocked the movement of fluorescent dye between cells. Preimmune serum or extracellular immune serum had no effect. In order to introduce these antibodies into many cells at once, hydras were treated with 5% DMSO. After this treatment, the hydras remained healthy. Hydra which had been exposed to immune sera in this manner displayed uncoordinated movement, consistent with a disruption of coupling between contractile cells. A small ring of apical tissue was grafted midway down the hydra body column. In normal hydra, the head suppressed the formation of a second head. If the head was removed, a second head formed. Treatment with immune sera significantly increased the incidence of head formation from the grafted tissue in the presence of the original head.

These results indicate that gap junctional communication plays a role in head inhibition, presumably by permitting diffusion between cells of a small head inhibitor molecule.

♦Positional information that leads to pattern formation can be transmitted via several different routes. The current study demonstrates that gap junctions play a role in the transmission of a low molecular weight moiety that directly regulates pattern formation in hydra. The authors introduce into hydra cells antibodies to the major rat gap junction protein. They then demonstrate that the antibodies not only recognize hydra gap junction antigen but also interfere with cell-cell communication mediated by the gap junctions. In grafting operations, the rat anti-gap junction antibodies directly interfere with the head inhibition gradient. These results demonstrate for the first time a direct role for gap junctions in tissue pattern formation. *Joel M. Schindler*

Contact-Independent Polarization of the Cell Surface and Cortex of Free Sea Urchin Blastomeres

T. E. Schroeder
Dev. Biol., 125, 255—264, 1988 9-6

Sea urchin embryo blastomeres are polarized, such that microvilli and pigment granules face the hyaline layer and do not face adjacent cells or the blastocoel. To test the role of cell contacts in the establishment of this polarity, fertilized sea urchin eggs were stripped of the hyaline layer and cultured to the 16-cell stage in calcium-free artificial seawater to prevent intercellular adhesion.

The free blastomeres created by this treatment divided normally. When examined by scanning electron microscopy, these blastomeres had developed polarization of microvilli and pigment granules. This self-polarization was confirmed by NBD-phallacidin stain of polymerized actin bundles associated with the polarized microvilli. Treatment with the cell division inhibitor colchicine uncoupled pigment granule polarization, which was normal, from microvilli polarization, which did not occur.

These results indicate that sea urchin blastomeres can self-polarize. Therefore, contact with the hyaline membrane and cell-cell contact are unnecessary for blastomere polarization. The nonrandom polarization indicates that there is coordination between the cleavage planes and the axes of polarization.

♦This study examines the mechanisms of determining the apical-basal polarity in early blastomeres of the sea urchin embryo. This is an important feature of early embryogenesis since the blastocoel begins to form during early cleavage divisions, and because cell cleavages occur in both an equatorial and longitudinal axis relative to the original orientation of the egg. Early observations (*Dan. Int. Rev. Cytol.,* 9, 321—326, 1960) demonstrate that blastomeres can become polarized, even when isolated from early cleavage stages. Yet other evidence suggests that certain contacts between cells and the extracellular matrix are responsible for blastomere polarity (Wolpert and Mercer, *Exp. Cell Res.,* 30, 280—300, 1963). In the present report, Schroeder removed the extracellular hyalin layer and isolated blastomeres from two cell staged embryos, and observed the formation and retention of apical/basal polarity in cells through the 16-cell stage. These observations are important because they indicate that some kind of coordination exists between the changing planes of successive cellular cleavages and the axes of autonomous polarizations. However, the polarity of cells from older embryos do appear to require some contact (Nelson and McClay, *J. Cell Biol.,* 103, 371a, 1987), so that some change must occur during development in the establishment and retention of cellular polarity. *Gary M. Wessel*

Determination of Dorso-Ventral Axis in Early Embryos of the Sea Urchin, *Hemicentrotus pulcherrimus*

T. Kominami

Dev. Biol., 127, 187—196, 1988 9-7

The organization of the dorsoventral (DV) axis in sea urchin embryos is not understood. To examine the relationship between early cleavages and the position of the DV axis, Lucifer Yellow CH was iontophoretically introduced into one cell of 2-cell stage *Hemicentrotus pulcherriumus* embryos and the location of progeny cells was determined from the animal pole.

Eight different labeling patterns were seen at the prism larva stage, corresponding to eight positions around the DV axis. The boundary that divided the embryo into labeled and unlabeled regions was coincident with, perpendicular to, or obliquely crossed the plane of bilateral symmetry.

This study demonstrates that Lucifer Yellow is useful as a tracer dye in early sea urchin development. The results suggest that the DV axis of *H. pulcherrimus* embryos is determined during the stage when the blastomeres align in eight rows in the meridional direction, in the fifth to sixth cleavage (32 to 64 cell) stage.

♦While much information exists on the animal-vegetal axis in sea urchin development, little is known about the other major embryonic axis, the dorsal-ventral (or oral-aboral) axis. This axis is of particular interest because genes now have been isolated that are expressed exclusively in aboral ectoderm cells (cf. Wilt, *Development,* 100, 559—575, 1987) and thus serve as molecular markers for the differentiation of this cell type. The establishment of the dorsal-ventral axis is believed to occur during the early cleavage stages but it is not certain at precisely which stage the axis is actually specified. In this article, the author attempts to answer this question by determining the relationship of the first cleavage plane, which occurs along the animal-vegetal axis, with the dorsal-ventral axis. Using Lucifer Yellow, iontophoretically introduced into a single cell of the two-cell stage *H. pulcherrimus* embryo, eight different labeling patterns are observed at the prism stage which correspond to eight discrete positions around the animal-vegetal axis. Parallel, perpendicular, and oblique patterns are seen with respect to the dorsal-ventral axis. The hypothesis that best fits these data is that the dorsal-ventral axis is specified during a stage when the blastomeres are aligned in eight rows in the meridonal direction and not before. This would be consistent with the 32- or 64-cell stage. An earlier labeling study carried out by Cameron et al. (*Genes Dev.,* 1, 75—84, 1987) using *S. purpuratus* embryos labeled at the eight-cell stage showed only six labeling patterns. These investigators found that progeny of single blastomeres of the eight cell stage embryo could give rise solely to oral or aboral cells, suggesting that the axis is specified at or earlier than the third cleavage. Based on their observations, Cameron et al. concluded that the oral-aboral axis probably lies at a 45° angle from the first and second cleavage furrows. There is no explanation for the discrepancy of the earlier results with the present study, other than the possibility of species differences, that is, in *S. purpuratus* the dorsal-ventral axis is specified earlier than in *H. pulcherrimus. William H. Klein*

Dorsal, an Embryonic Polarity Gene in _Drosophila_, Is Homologous to the Vertebrate Proto-oncogene, _c-rel_

R. Steward

Science, 238, 692—694, 1987

9-8

In *Drosophila,* the establishment of the dorsal-ventral axis is dependent upon both maternal effect and zygotic genes. *Dorsal* is one of 11 such maternal effect loci. Females homozygous for this mutation produce embryos with severe disturbances of the dorsal-ventral axis. This gene appears to be active in early embryogenesis before cellular blastoderm formation (2.5 h). Recently, *dorsal* has been cloned and characterized, and this report describes its sequence.

The *dorsal* gene encodes a protein of 677 amino acids with a molecular weight of 75,600. This protein was found, by computer search, to contain a stretch of 295 amino acids which is 47% identical to the avian protooncogene c-*rel* and to the corresponding transforming gene of the reticuloendotheliosis virus v-*rel.* (This retrovirus is highly oncogenic in chicks and poults, and can transform spleen cells *in vitro.)* When conservative amino acid changes are scored positive, the homology is about 80%. There is a human c-*rel* gene homologous to the avian c-*rel,* and *dorsal* has about the same similarity to it as to the avian gene. The homologous region contains one possible serine phosphorylation site and a possible glycosylation site. The nonhomologous portion of the predicted protein contains stretches of single amino acids similar to those that have been found in other *Drosophila* proteins.

These data suggest that *dorsal* and *rel* may have similar functions. Little is known about c-*rel* except that cross-hybridizing RNA has been found in spleen, muscle, and hematopoietic tissue. *Dorsal* may function in the nucleus, and the author suggests that it might differentially activate zygotic gene expression to give rise to dorsal-ventral asymmetry.

♦At about 3 h of development the *Drosophila* embryo consists of an ovoid monolayer of morphologically identical cells. Mutational analysis has identified a set of a dozen or so genes that directly affect determination of these cells with respect to the dorsal-ventral axis. In mutants of this class, ventral cells assume more dorsal fates, while determination along the anterior-posterior axis is not affected. The gene *dorsal* is believed to act at the end of the "cascade" of this group or dorsalizing genes and may correspond to the actual effector gene. This paper reports the sequence of a *dorsal* complementary DNA clone. The predicted protein has 677 amino acids and a molecular mass of 75,600 kDa. The most significant finding was that the predicted protein is homologous over a 295-amino-acid stretch with the avian oncogene *v-rel* carried by the highly oncogenic reticuloendotheliosis virus that infects chicks and poults. The homology between *dorsal* and *v-rel* is of about 47%, or 80% if one includes conservative amino acid changes. Recently, several genes that play an important role in embryonic determination have been shown to have homology to oncogenes. By comparison with some other *Drosophila* developmentally important genes, the structure of *dorsal* is relatively simple, a property that should facilitate further studies. The realization that the function of the cellular homologs of oncogenes is beginning to be elucidated is exciting. *Marcelo Jacobs-Lorena*

The *Toll* Gene of Drosophila, Required for Dorsal-Ventral Embryonic Polarity, Appears to Encode a Transmembrane Protein

C. Hashimoto, K. L. Hudson, and K. V. Anderson
Cell, 52, 269—279, 1988 9-9

The *Toll* gene of *Drosophila* is one of 11 maternal effect genes comprising the dorsal-group genes. Loss-of-function mutations in any of these results in dorsalized embryos lacking all lateral and ventral structures. For *Toll,* there exist dominant gain-of-function allelles which cause ventralization of embryos. This paper presents the results of the isolation and characterization of the *Toll* gene.

Toll was cloned using P element insertions and a screen for the phenotypic reversion of a dominant *Toll* allele. The genomic clone was used to hybrid select mRNA. The injection of a 5.3-kb transcript so purified from the ovaries of wild-type flies rescued embryos from *Toll*⁻ mothers. The genomic clone was then sequenced, and an open reading frame was found which encoded a protein of 1097 amino acids with a molecular weight of 125,000. The primary structure of this protein suggested that it is a transmembrane protein with a cytoplasmic domain of 269 amino acids and a larger extracellular domain of 803 amino acids. The latter domain contained 17 potential glycosylation sites. A computer search found no proteins similar to the cytoplasmic domain of *Toll,* but showed that 15 repeats in the extracytoplasmic domain of a 22- to 26-amino acid leucine-rich segment was similar to a repeat in the leucine-rich α_2-glycoprotein (LRG), a minor human serum protein of unknown function. Similar sequences are found in yeast adenylate cyclase, *Drosophila* chaoptin, and the proteoglycan core protein of human fibroblasts.

The authors suggest that if *Toll* is indeed a membrane protein and since *Toll* functions at the syncytial blastoderm stage of development, it might be localized either on a membrane facing the outside of the embryo or on an intracellular membrane. Further progress on the function of the *Toll* protein will therefore require knowledge of its localization.

♦Pattern formation is the result of cells responding to positional information. In *Drosophila,* there are many known genes that are required for accurate antero-posterior and dorso-ventral development. It is likely that some of the products encoded by such genes convey positional information. One such gene is the *toll* gene. The authors report here the cloning of the *toll* gene. Subsequent sequence analysis suggests that the gene product could be a transmembrane protein. As such, it suggests that some type of cell interaction may be involved in transmitting the positional information needed for normal dorsal-ventral polarity. *Joel M. Schindler*

Identification and Spatial Distribution of Retinoids in the Developing Chick Limb Bud

C. Thaller and G. Eichele
Nature (London), 327, 625—628, 1987 9-10

It has long been suspected that pattern formation is due to local concentration gradients, but the molecules responsible, or morphogens, have proven difficult to identify. In the developing chick, it is known that posterior limb bud tissue (the zone of polarizing activity, ZPA) can induce digit pattern duplications when grafted anteriorly, and recent experiments have shown that this effect can be mimicked by all-*trans*-retinoic acid (RA). This resemblance has led to the hypothesis that RA is related to the inducer believed to be released by the ZPA. The experiments in this paper were performed in a search for endogenous RA in chick limb buds.

By several lines of evidence, RA was extracted and identified from limb buds of stage-21 chick embryos (see Figure 9-10). Reverse phase HPLC analysis showed a peak of UV-absorbing material comigrating with ^{3}H-RA that was added as an internal standard before tissue extraction. Repurification of this peak by normal phase chromatography gave a single peak coeluting with the standard. Both UV-visible and fluorescence spectra of

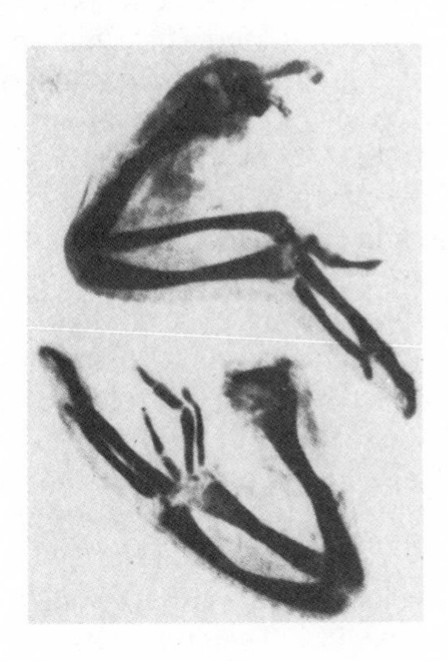

FIGURE 9-10. Chick wing with a *43234* digit pattern that developed following local application of RA extracted from stage-21 chick embryos. The specimen above is the contralateral, untreated control wing *(234* pattern). Methods: RA was extracted from 600 stage-21 embryos and purified by reverse and normal phase HPLC as described for limb buds. RA (117 ng) was dissolved in 5 µl DMSO to a final concentration of 23 µg/ml. Twelve AG1-X2 beads (in formate form, Bio-Rad) were soaked in this solution for 30 min. Each bead was briefly rinsed in phosphate-buffered saline and implanted at the anterior margin of a wing bud of a stage-20 embryo. The embryo was incubated for six additional days, killed, and stained with alcian green, a cartilage dye. (Reprinted by permission from *Nature*, 327, 627. Copyright © 1987 Macmillan Magazines Ltd.)

this single peak were identical to that of authentic RA. The extracted material was found to be as efficient as synthetic RA in a bioassay for pattern duplicating ability. Finally, RA was quantified for anterior and posterior segments of limb buds. Both portions contained the same amount, but when normalized for either DNA, protein or lipid content, a two- to threefold enrichment of the posterior region was revealed.

The authors provide evidence that this modest concentration gradient is similar to that established by exogenously applied, bioactive RA. The fact that RA forms a concentration gradient in the developing limb bud lends support to the view that RA is a morphogen.

♦During development, tissues and body parts form in characteristic patterns according to a strict developmental program. This process of "pattern formation" is striking in its preciseness and reproducibility. Limb development represents a prominent model system for studying pattern formation, since the limb develops from a paddle-shaped limb bud into a complex structure with proximo-distal, anterior-posterior, and dorso-ventral polarity. Along the anterior-posterior axis of the limb, the digits form in a characteristic pattern such that the "thumb" is anterior and the "pinky" is posterior. What controls this polarity? One model is that a "zone of polarizing activity" (ZPA) exists in the posterior portion of the limb bud. The ZPA has been proposed to be the source of a "morphogen" which diffuses radially outward, creating a gradient of a diffusible substance which causes posteriorization. Although the morphogen was not identified, it has been well established that retinoic acid can also cause posteriorization of the limb, thus mimicking the effects of the proposed morphogen.

In this study, Thaller and Eichele have examined limb bud extracts for retinoic acid using the HPLC. They found that retinoic acid is indeed present in the limb bud and appears to be graded, with higher amounts in the posterior third than in corresponding volumes of anterior limb buds. This provides the first direct evidence that retinoic acid is present in limb buds and therefore may actually be the morphogen made in the ZPA. This work is significant because it represents one of the few examples where a morphogen with demonstrated biological effects has been isolated from embryonic tissues.

CAVEAT: The criticism of this work is that the authors are working at the limit of detectability of the HPLC. Therefore, a twofold difference between posterior and anterior quadrants is not overwhelmingly convincing. Still, there is a lot of excitement over this work, so I chose to include it. *Marianne Bronner-Fraser*

The Effect of Somite Manipulation on the Development of Motoneuron Projection Patterns in the Embryonic Chick Hindlimb

C. Lance-Jones
Dev. Biol., 126, 408—419, 1988 9-11

A variety of observations of normal and experimentally perturbed development of chick motoneurons suggest that projections from the eight lumbosacral (LS) cord segments to limb muscles involve specific recognition. One hypothesis for the mechanism of this recognition is that early in development there is joint labeling of motoneuron and muscle cell precursors at the same axial level, and that the formation of projections to the correct muscle involves label matching. As tests of this hypothesis, previous experiments have shown that when three to four anterior LS cord segments were reversed before the development of motoneuron outgrowth, normal axonal projections resulted. The experiments described here were designed as complements to these experiments.

Somites of levels 26—29 or 27—29 were rotated about their anteroposterior axis in the period before motoneuron projection. This will result in reversal of somite origin of anterior thigh muscles. When projections to these muscles were followed by retrograde HRP labeling, largely normal results were evident. A second series of experiments was then undertaken in which somites plus adjacent spinal cord segments at levels 26—29 were both reversed. Here again, the muscles formed normally, but the projections of the motoneurons were the reverse of normal.

The first series of experiments suggests that muscle cells or somites do not bear cues for specific recognition by motoneurons. The second series suggests that neither are projections of motoneurons guided by cues in the neural tube. The author concludes that either cues for motoneuron outgrowth are associated with embryonic tissues other than somites or spinal cord, or cues become associated with myogenic cells only after motoneuron projections begin.

♦During development, there is a stereotypical innervation pattern of motoneurons and their target muscle. In the hind-limb, Lance-Jones has shown that motoneurons from a given segmental level project to muscles derived from somites of that same axial level. These data suggest that positional cues influence the patterning of both muscles and their neuronal connections. When somites are transplanted from one region to another, they form muscles appropriate to the region where they were grafted. This suggests that environmental cues characteristic of each axial level guide muscle cell precursors to appropriate muscles.

In this study, Lance-Jones rotates the somites or the neural tube plus the somites about their anterior-posterior axis. After somite rotation, muscles form normally as if no rotation had been performed. This is consistent with the hypothesis that the environment guides muscle cell precursor to their final sites. When both the neural tube and somites are rotated, the muscles again form normally. However, the motoneurons connect to the muscle appropriate to their axial level of origin. The environment then appear to "guide" axons to the appropriate muscle. This study is highly informative about the nature of the cues governing neuronal connections in the periphery. First, it is likely that the guiding mechanisms are determined by environmental cues that are outside of the neural tube and somites. These may reside within the other embryonic tissues such as connective tissue. Second, it seems likely that motoneurons are already specified at the time of the graft to project to certain muscle groups and thus have a positional identity prior to rotation. Thus, neuronal patterning may depend upon both inherent positional information in the neurons and environmental guidance of muscle precursors and neurite to appropriate sites. *Marianne Bronner-Fraser*

Proximal Tissues and Patterned Neurite Outgrowth at the Lumbosacral Level of the Chick Embryo: Deletion of the Dermamyotome

K. W. Tosney

Dev. Biol., 122, 540—558, 1987 9-12

In the chick limb, the projection of spinal nerves to their target muscles is guided in both a specific and a permissive sense, and somites are obvious candidates for guiding tissues. Spinal nerves are segmented, and this has long been believed to be secondary to the myogenic somites. Just before axon outgrowth, each somite differentiates into two histologically distinct portions: the dorsal dermamyotome, precursor to skeletal muscle and dermis, and the ventral sclerotome, precursor to vertebral cartilage. The experiments reported here were designed to investigate the role of the dermamyotome in the development of neurons.

A single dermamyotome was removed from 35 chick embryos, and 2 or more adjacent dermamyotomes were removed from 27. Operations took place at three stage groups: before axon outgrowth, during early outgrowth, and as axons reach the plexus region. Neurite projection patterns were analyzed by HRP injection or by sectioning and staining. Extirpation of dermamyotome had no apparent effect on migration or condensation of neural crest cells that form sensory ganglia, on spinal nerve organization, or on motor nerve sorting and targeting in the limb. However, after dermamyotome(s) deletion, the corresponding epaxial muscles were absent and the dorsal ramus innervated the epaxial muscle in the closest adjacent segment. When dermamyotomes in adjacent segments were removed or reduced, no dorsal ramus formed.

These results suggest but do not prove that the dermamyotome has no effect on various patterning events suspected to depend on it. The surprising finding that the outgrowth of the dorsal ramus depended upon dermamyotome suggests that its development normally proceeds via chemotactic cues provided by the dermamyotome.

♦The projection of spinal nerves to their target muscle in the limb is highly patterned and reproducible. Both motor axons and neural crest cells, which form the dorsal root ganglia, migrate through the anterior half of each somite. Because they are absent from the posterior half of each somite, this leads to an apparently metameric distribution pattern along the rostro-caudal axis. At the stages of motoneuron outgrowth, the somite consists of the dermamyotome and the sclerotome. Some controversy existed about the relative contributions of these components to the patterning of axons and neural crest cells.

In this investigation, Tosney has surgically removed the dermamyotome and examined the subsequent effects on the formation of neural crest-derived dorsal root ganglia and on the patterning of spinal nerves. In most cases, the ganglia, spinal nerves, and projections to the limb occurred perfectly normally. These results suggest that the sclerotome, which remains intact after the operation, provides the cues that cause proper segmentation of ganglia and nerves. The single defect observed was within the dorsal ramus which projects to the dermamyotome under normal circumstances. In the absence of its target, the dorsal ramus either was absent or sent its projections to adjacent segments. These are important results, first, because they resolve a controversy over whether the dermamyotome or sclerotome controls the segmental pattern of neural crest migration and motoneuron processes. It is clear that the dermamyotome is not necessary for proper segmentation of the nervous system. Second, these results suggest that the dermamyotome provides a chemotactic signal for dorsal ramus axons. Since few chemotactic signals have been described in vertebrates, this opens an interesting area of research. *Marianne Bronner-Fraser*

Altered Proteoglycan Synthesis Disrupts Feather Pattern Formation in Chick Embryonic Skin

P. F. Goetinck and D. L. Carlone

Dev. Biol., 127, 179—186, 1988 9-13

It has previously been suggested that proteoglycans are involved in skin development. The present study tested this possibility by altering the structure of proteoglycans in developing chick skin and assessing the results on feather patterns.

Skin segments from 7-d-old chick embryos were cultured in the presence of either 2 m*M para*-nitrophenyl-β-D-galactoside or control medium. Xylosides act as exogenous acceptors for the synthesis of glycosaminoglycans and therefore compete with core proteins of proteoglycans. This competition leads to the synthesis of underglycosylated core proteins and free glycosaminoglycans. When skins were cultured for 24 h in the presence of $^{35}S[H_2SO_4]$, abnormal proteoglycans were synthesized, as demonstrated by both gel filtration and density gradient assays. At the same time, abnormal feather rudiment patterns developed, with feather rudiments fused in diagonal, anteroposterior, or mediolateral directions. This disturbance of feather pattern occurred when skins were incubated with xyloside in the first day in culture, but this effect was reversible if skins were returned to control medium for the next 2 d. Once a normal feather pattern was established, however, subsequent treatment with xyloside merely reduced the size of the individual feather germs, but did not disturb the pattern.

These experiments have shown that disruption of embryonic skin proteoglycans results in alteration of feather patterns. Because xyloside treatment affects the establishment but not the maintenance of this pattern, this work suggests that these two processes are under different developmental controls.

♦Proteoglycans are extracellular matrix molecules composed of a protein core linked to numerous unbranched glycosaminoglycan chains. Although proteoglycans have been found in abundant locations within the embryo, the role of these molecules in morphogenesis and pattern formation remains largely unexplored.

In this study, Goetinck and Carlone have examined the effects of an inhibitor, B-xyloside, on development of avian feather pattern in the skin. B-xyloside interferes with glycosylation of the proteoglycan core proteins, resulting in core proteins and free glycosaminoglycan chains. The authors find that skin treated with B-xyloside exhibits abnormalities in the feather pattern manifested by fusion of the feather rudiments. Most interestingly, the skin is sensitive to B-xyloside during the time of initial pattern formation, and the effects of the drug are reversible for a short period of time. These results suggest a role for proteoglycans in the establishment of feather pattern. This represents one of the first instances where proteoglycans have been definitely shown to alter morphogenesis. However, these molecules do not appear to be necessary for maintenance of feather pattern, indicating that proteoglycans play a transient role in the skin. Other molecules, like L-CAM may also be involved in this pattern event, highlighting the possibility that multiple molecules acting cooperatively or independently may influence morphogenetic processes. *Marianne Bronner-Fraser*

Effect of the Notochord on the Differentiation of a Floor Plate Area in the Neural Tube of the Chick Embryo

H. M. W. van Straaten, J. W. M. Hekking, E. J. L. M. Wiertz-Hoessels, F. Thors, and J. Drukker

Anat. Embryol., 177, 317—324, 1988 9-14

Previous studies in both amphibians and chicks have led to the suggestion that the notochord induces the floor plate in the neural tube. These studies included the induction of an additional floor plate-like area by a notochord fragment implanted adjacent to the neural tube. In this work, several quantitative and qualitative measures were used to define the morphology of an induced floor plate and to determine the critical period of its induction.

Chick embryos 1.5 to 2 d old were implanted with either a notochordal segment or an inert spacer (fragment of a human hair). The implant occurred adjacent to the lateral wall of the neural groove between prospective wing and leg buds. In 16 of 17 operated embryos, an additional floor plate developed facing the implanted notochord. This additional floor plate resembled the natural floor plate by various criteria including its morphology, its relatively thin neural tube wall, its elongated nuclei, its low number of mitotic figures, and its apparent absence of neuroblasts. The inert spacer had no effect. The induction occurred in 1.5-d-old embryos, but not in those 2 d old.

The authors suggest that the notochord induces floor plate cells by reducing the proliferative ability of neural plate cells and blocking their ability to differentiate into neuroblasts. Because the notochord can also stimulate proliferation of neural plate cells at later times in development, studies on the effect of an additional notochord on cell cycle and proliferation are in progress.

♦The notochord forms from the chordomesoderm which comes to lie under the presumptive neural plate just after gastrulation. One important function ascribed to the notochord is that of neural induction. In support of this, transplantation of an extra notochord underneath the ectoderm results in two nervous systems.

The neural tube, which gives rise to the entire central nervous system comprised of the brain and spinal cord, has a characteristic dorso-ventral polarity. Shortly after neural tube formation, neural crest cells emerge from the dorsal neural tube and motor neuron processes emerge from the ventral neural tube. Later, the neural tube has distinct dorsal and ventral horns. One characteristic of the ventral neural tube is a morphologically distinct region called the floorplate, which lies in the ventral midline.

In this study, van Straaten and colleagues examine the effects of the notochord on formation of the floorplate region. They find that implantation of an ectopic notochord adjacent to the neural tube results in the formation of an additional floorplate. However, the neural tube will only respond to this notochordal induction for a limited period of time during its development. This work has very important implications regarding the factors controlling dorso-ventral polarity of the nervous system. Since the notochord apparently induces the ventral-most aspect of the neural tube (i.e., the floorplate), it may be responsible for establishing the dorso-ventral orientation of the spinal cord. *Marianne Bronner-Fraser*

Pax 1, A Member of a Paired Box Homologous Murine Gene Family, Is Expressed in Segmented Structures during Development

U. Duetsch, G. R. Dressler, and P. Gruss
Cell, 53, 617—625, 1988 9-15

A number of different gene sequences important in morphogenesis in *Drosophila* have been found in phylogenetically distant species. These included homeoboxes, "finger" domains, and oncogenes. This paper presents the results of an investigation into the possible conservation in mouse of the paired box sequence *(Pax)*, found in a subset of *Drosophila* segmentation genes consisting of *paired, gooseberry proximal,* and *gooseberry distal.*

Mouse genomic libraries were screened with a probe containing the *Pax* sequence from the *gooseberry distal* gene of *Drosophila.* Several homologous mouse sequences were detected, indicating the existence of a multigene family. The nucleic acid sequence of a mouse *Pax*-containing gene fragment *(Pax-1)* was determined, and it proved to be 58 to 63% identical to the three *Pax* sequences of *Drosophila;* the deduced amino acid sequence was 64 to 68% identical to the *Drosophila* sequences. Northern blot analysis showed a single cross-hybridizing 3.1-kb transcript present in mouse embryos, with peak concentrations present in 12- to 14-d-old embryos. This transcript was not found in any of eight adult tissues tested, including brain. *In situ* hybridization to frozen sections of mouse embryos of various stages suggested that at 10 d of development, ventral sclerotome cells, which express *Pax-1,* migrate to the notochord at the intervertebral disk anlagen of the perichordal zone. *Pax-1* transcripts were also detected in the thymus of 12-d-old embryos, and the third and fourth pharyngeal pouches of 10-d embryos.

The expression of *Pax-1* appeared first as an uninterrupted line along the rostral-caudal axis, and later only in condensed perichordal tissue. The latter type of expression, in a "striped" pattern, resembles expression of segmentation genes. In contrast to homeobox-containing genes that appear to specify positional information, *Pax-1* may be required for describing the entire rostrocaudal axis.

♦The *paired box* sequences represent a conserved domain (128 amino acids) found in *paired, gooseberry proximal,* and *gooseberry distal,* all of which are *Drosophila* segmentation genes of either the pair-rule or segment polarity classes (Bopp et al., *Cell,* 47, 1033—1040). The *gooseberry distal paired box* nucleotide sequence was used to detect several conserved sequences in the mouse genome by Southern blot analysis, indicating the presence of a multigene family. By screening a genomic library, a mouse *paired box* (*Pax 1*) sequence was isolated. This particular sequence showed extensive nucleotide (60%) and amino acid sequence (approximately 70%) conservation to the three *Drosophila Pax* sequences. No homology flanking the box was found. The mouse *Pax 1* gene detected a 3.1-kb transcript present in embryos (day 10 to 17). Transcripts were not detected in adult tissues (brain, heart, kidney, liver, lung, ovary, spleen, or testis). *In situ* analysis revealed hybridization to sclerotome cells of the differentiated somite (days 9 and 10), ventral mesenchyme directly lateral to the notochord (day 10), and the perichordal zone of the developing vertebral column (day 12 to 14). *Pax 1* transcripts were also detected in thymus (day 12) and the third and fourth pharyngeal pouches (day 10).

Work of other investigators have described the expression of homeobox genes (another conserved domain that is different from the *Pax* sequences, see Homeobox summary in this volume) as being region specific, thus establishing positional domains along the anterior-posterior axis. In contrast, *Pax 1* was found to be expressed along the entire rostro-caudal axis independent of position. Although this pattern of expression suggests a role important for establishing the entire axis, the definitive role of *Pax 1* remains to be determined. It will be of interest to determine the pattern of expression of the other *Pax* genes that have been identified but not yet analyzed. Antibodies to the gene products will be important for functional analysis as will mutations in the gene.

Until recently , it has not been possible to identify genes in mouse that have potential key roles involved in the genetic control of mammalian development. The use of conserved sequences (homeobox and *Pax*) in *Drosophila* to identify genes expressed in defined patterns and tissue types during mouse development is exciting and will undoubtedly add to the understanding of murine development. Although other *Drosophila* sequences have been tested for sequence conservation in the mouse genome, those showing negative results (e.g., *sex-lethal*) have probably not been reported. It would be extremely beneficial to investigators studying mammalian development to have at their disposal a catalogue, perhaps published in *Mouse News Letter,* of those *Drosophila* sequences that have been examined and whether they yielded positive or negative results. *Terry Magnuson*

A Human Retinoic Acid Receptor which Belongs to the Family of Nuclear Receptors

M. Petkovich, N. J. Brand, A. Krust, and P. Chambon
Nature (London), 330, 444—450, 1987 9-16

Vitamin A (retinol) and its derivatives have been shown to play crucial roles in differentiation and development, and much evidence suggests that retinoic acid (RA) is the endogenous morphogen giving positional information in the chick limb bud. Much evidence has suggested that RA may, like steroid hormones, act via specific nuclear receptors. Based on the striking conservation of the sequence of the DNA binding region of all known members of the nuclear (steroid) receptor multigene family, the experiments reported here were undertaken in order to screen cDNA libraries from RA-sensitive breast cancer cells for the putative RA receptor gene.

A probe was prepared corresponding to a consensus sequence from six members of the nuclear receptor family. A clone was identified which had 21 of 24 residues of the consensus sequence, which contained an open reading frame of 1296 nucleotides encoding a protein with a molecular weight of 48,000. It resembles other members of the nuclear receptor family, particularly the thyroid hormone receptor hc-erbAβ. The encoded protein binds RA with high affinity and specificity. Cells transfected with a chimeric gene, containing the hormone-binding domain of this retinoic acid receptor (RAR) gene and the estrogen receptor gene's enhancer binding domain, showed an increase in reporter gene activity when stimulated with RA concentrations 10^{-9} *M* or greater. This concentration corresponds to that necessary for many known biochemical and developmental effects of RA. Northern analysis using hRAR cDNA as probe showed cross-hybridization to several mouse tissues.

This putative RAR gene is also closely related to a genomic sequence of a hepatoma into which the hepatitis B virus is inserted. This similarity raises the possibility that mutations in RAR might be oncogenic, a reasonable possibility in light of the myriad developmental effects of vitamin A and its derivatives.

Identification of a Receptor for the Morphogen Retinoic Acid
V. Giguere, E. S. Ong, P. Segui, and R. M. Evans
Nature (London), 330, 624—629, 1987 9-17

The work reported here is part of a series in which the conserved domain of the nuclear hormone receptor gene family was used as a probe to scan the genome for related receptors. Previously identified members of this family have helped elucidate the mechanisms mediating gene regulation in response to inducers like hormones and growth factors. In this paper, a new cDNA of this multigene family is described.

The clone isolated from a λgt10 kidney cDNA library contained an open reading frame of 462 amino acids or approximately 51 kDa. Several lines of evidence are presented which suggest that the clone described here is a human retinoic-acid receptor (hRR). First, it is related to steroid and thyroid receptors, most strongly to the latter. Second, retinoic acid induced expression of a reporter gene when COS-1 cells were transfected with a chimeric receptor containing the DNA binding domain of the human glucocorticoid receptor and the putative ligand-binding domain of hRR. This induction occurred at concentrations of retinoic acid active in several bioassays. Retinol was a weak agonists. Also, COS-1 cells transfected with an expression vector containing the putative hRR gene showed increased binding capacity for retinoic acid. Northern blot analysis found cross-hybridizing RNA present in most rat and human tissues tested, with high levels in brain, adrenals, and testis, and undetectable levels in liver.

Southern blot analysis suggested that a family or RAR-related genes exist. The authors propose that the members of this family might be specific receptors for different vitamin A derivatives.

♦Attempts to identify the chemical nature of putative morphogens has been an ongoing research effort. Recently, retinoic acid was shown to be a true morphogen, and investigators have attempted to explain its mechanism of action. Initially, a cytoplasmic binding protein was identified and implicated as being functionally significant. However, the papers by Petkovich et al. and Giguere et al. identify a retinoic acid receptor unrelated to the binding protein and part of a multigene family. This receptor has strong molecular homology to steroid receptors and is directly implicated in the mechanism of action of retinoic acid. Transfection experiments indicate that this molecule can selectively increase the ability of cells to bind retinoic acid. This is the first identification of a molecule directly involved in the mode of action of a known morphogen. *Joel M. Schindler*

AUTHOR INDEX

A

Aerts, R. J., 52
Aizawa, S., 68
Akhurst, R. J., 8
Albert, I., 122
Alexander, S., 4—6, 8, 30—32, 53, 54, 86, 88, 89, 146, 202—205
Allen, J. M., 44
Allen, N. D., 45
Allison, J., 167
Alpert, S., 143
Amaldi, I., 48
Anderson, K. V., 210
Anderson, P., 10
Andrews, M. E., 55
Angerer, L. M., 148
Angerer, R. C., 148
Anstrom, J. A., 147
Artavanis-Tsakonas, S., 56
Austin, J., 134
Avery, L., 137
Awgulewitsch, A., 192, 197

B

Bailes, J. A., 173
Baker, B. S., 156
Ballard, D. W., 66
Balling, R., 81
Baltimore, D., 23
Banerjee, U., 94, 95
Barbas, J. A., 120
Barberis, A., 33
Barde, Y.-A., 164
Bargiello, T. A., 101
Baron, A., 190
Barstead, R. J., 151
Barton, S. C., 45, 80
Basler, K., 93
Bateman, J., 22
Baylies, M. K., 101
Behringer, R. R., 44
Bell, J. I., 26
Belote, J. M., 155, 156
Benian, G. M., 151
Bennett, W. S., 125, 126
Benson, S., 131, 132
Benz, W. K., 13
Benzer, S., 94, 95
Berks, M., 203
Berleth, T., 76
Birnsteil, M. L., 37
Bixby, J. L., 113
Bjorkman, P. J., 125, 126
Bode, H. R., 205
Boetiger, D., 111
Boggs, R. T., 155, 156
Bohnlein, E., 66
Bolton, V., 49
Bonhoeffer, F., 110, 114
Bopp, D., 76
Bowtell, D. D. L., 95, 96
Bozzaro, S., 88
Bradley, A., 19
Brand, N. J., 219
Brandhurst, B. P., 34
Braude, P., 49
Bravo, R., 58
Breen, E. J., 204
Bremer, K. A., 57
Brinster, R. L., 44, 117, 167
Britten, R. J., 8, 39, 132, 149
Bronner-Fraser, M., 66, 105, 106, 108—112, 114—117, 163—165, 212, 214—217
Brown, D. D., 43
Brown, L. G., 170
Brulet, P., 195
Buehr, M., 169
Burgoyne, P. S., 169
Burkly, L. C., 117, 167
Burne, J. F., 108
Burns, R. A., 29
Burri, M., 76
Burtenshaw, M. D., 196
Burtis, K. C., 156
Busslinger, M., 33, 37
Byrne, G., 201

C

Cameron, R. A., 149
Campbell, I. L., 167
Campos-Ortega, J. A., 57, 100
Capecchi, M. R, 16
Carlone, D. L., 215
Carroll, S. B., 181
Catlin, T. L., 38
Ceccarelli, , A., 203

Chaix, J.-C., 120
Chalfie, M., 91, 180
Chambon, P., 219
Chang, S., 110
Chapelle, A. de la, 170
Chapman, N., 129
Cheng, N., 71
Chien, Y.-H., 119
Childs, G., 35
Chin, J. E., 147
Cho, A., 86
Choi, A. H. C., 86
Choi, T., 22
Clayton, D. F., 162
Cohen, J., 108
Colberg-Poley, A. M., 192
Cole, W., 22
Collick, A., 80
Collin, A. M., 38
Collins, J., 10
Coltey, M., 116
Condamine, H., 21, 195
Conlon, R. A., 34
Cooke, J., 60
Cooley, L., 12
Corbel, C., 116
Cowing, C., 167
Cox, E. C., 201
Cran, D. G., 45
Crawford, D. R., 78
Crawley, C. R., 79
Crews, S. T., 158, 159
Culotti, J. G., 9—12, 72—75, 92, 93, 135, 137—140, 151—153, 181
Cunningham, B. A., 115

D

Dabauvalle, M. C., 58
Danielpour, D., 64
Dart, L. L., 64
Davidson, E. H., 8, 39, 131, 132, 149
Davis, C. A., 191
Davis, M. M., 119
Davis, R. L., 166
Dawid, I. B., 64
DeFay, K., 122
De Lozanne, A., 6
DeRobertis, E. M., 188
D'Eustachio, P., 196
Deutsch, W. A., 139
Dimond, R. L., 29
DiNardo, S., 42, 181
Dioree, M., 58
Doe, C. Q., 160, 161
Doetschman, T., 17
Dohrmann, U., 163
Dolecki, G. J., 179
Dony, C., 191
Dressler, G. R., 218
Drukker, J., 216
Duboule, D., 190
Duetsch, U., 218
Dustin, M. L., 123

E

Early, A. E., 146, 203
Edelman, G. M., 115
Edgar, D., 106, 163
Edgar, L. G., 150
Edstroem, J. E., 93
Eib, D. W., 14
Eichele, G., 211
Eldon, E. D., 148
Elliott, J. F., 119
Elmer, J. S., 10
Engels, W. R., 13
Ennis, H. L., 32
Ettensohn, C. A., 133
Evans, E. P., 196
Evans, M. J., 19
Evans, R. M., 220
Ewer, J., 101

F

Fainsod, A., 192
Farza, H., 81
Featherstone, M. S., 190
Feinberg, M. B., 127
Feng, Y., 30
Ferguson, E. L., 138
Fibi, M., 192
Finney, R., 4
Firtel, R. A., 51
Fisher, E. M. C., 170
Flavell, R. A., 167
Flavell, R. H., 117
Fleming, J., 10
Forhhofer, H. G., 75
Franks, R. R., 39, 149
Franza, B. R., 66
Fraser, S. E., 141, 205
Fregerio, G., 76
Fritz, A., 188

Frohman, M. A., 198
Fuller, D. A., 87
Fuller, M. T., 154

G

Gabert, J., 25
Gallin, W. J., 115
Galliot, B., 190
Gasic, G., 101
Gaunt, S. J., 193, 196
Gehring, W. J., 153, 160, 184, 185
Gelinas, R. E., 44
Gerisch, G., 88
Gibson, G., 185
Giebelhaus, D. H., 14
Giguere, V., 220
Gilula, N. B., 205
Giorda, R., 32
Gizang-Ginsberg, E., 198
Goetinck, P. F., 215
Gomer, R. H., 51
Goodman, C. S., 158—161
Gooley, A. A., 146
Goridis, C., 120
Grainger, R. M., 103
Green, C. R., 205
Greene, W. C., 66
Gregg, R. G., 17
Gregor, P., 155
Griscelli, C., 48
Gruss, P., 191, 192, 218
Gunderson, R. W., 107
Gurdon, J. B., 188

H

Hadam, M. R., 48
Hafen, E., 93, 95
Hall, A., 190
Hall, J. C., 101, 102
Handyside, A., 19
Hanrahan, D., 143
Hantaz-Ambroise, D., 104
Hardy, K., 19
Harkey, M. A., 147
Harmony, J. A. K., 25, 26, 28, 47, 49, 68, 119, 120, 123, 125, 127, 129, 130, 143, 169, 173
Harrison, L. C., 167
Hartley, D. A., 56
Harvey, R. P., 189
Hashimoto, C., 210
Hatchouel, M., 81
Heasman, J., 79
Heemskerk-Jongens, J., 42
Heimfeld, S., 142
Hekking, J. W. M., 216
Henderson, R. A., 103
Henry, J. J., 103
Herman, R. K., 9
Hettle, S., 45
Hill, D. P., 72
Hill, R. E., 190
Hinton, D. R., 95
Hiromi, Y., 160
Hodgkin, J., 152
Hofer, M. M., 164
Hogan, B. L. M., 194, 195
Holland, P. W. H., 194, 195
Holmann, H.-P., 88
Holwill, S., 79
Hooper, M. L., 17, 19
Horvitz, H. R., 137, 138
Hough-Evans, B. R., 39, 149
Hud, J., 114
Hudson, K. L., 210
Huecas, M. E., 162
Huey, T., 187
Humphreys, T., 179
Hunt, T., 54
Hunter, S., 19
Hursh, D. A., 55
Hyman, A. A., 74

I

Idriss, S., 155
Ito, K., 138

J

Jacob, F., 21
Jacobs-Lorena, M., 14, 41, 43, 58, 78, 99, 101, 103, 154, 155, 158, 160, 162, 184, 187, 188, 209
Jaenisch, R., 22
Jancarik, J., 171
Jermyn, K. A., 202, 203
Johnson, P. F., 46
Johnson-Schlitz, D. M., 13
Joyner, A. J., 191
Joyner, A. L., 194, 198
Judelson, H. S., 29

K

Kappler, J., 117

Karsenti, E., 58
Kassis, J. A., 42
Kay, R. R., 202, 203
Kelley, M. R., 139
Kelley, R., 12
Kemler, I., 37
Kemphues, K. J., 71
Kern-Veits, B., 114
Kessel, M., 192
Keynes, R. J., 65
Kidd, S., 139
Kidson, S. H., 38
Kiff, J. E., 11
Kim, S.-H., 171
Kimble, J., 134
Kimelman, D., 61
Kimmel, C. B., 140
King, W., 196
Kirschner, M., 61
Klein, W. H., 9, 34, 35, 37—39, 55, 148—150, 180, 208
Klemenz, R., 184
Knecht, D. A., 5, 87
Knust, E., 57
Koenig, J., 104
Kominami, T., 207
Komitowski, D., 175
Koopman, P., 169
Kosek, J. C., 127
Kraft, B., 4
Kriegler, M., 122
Krumlauf, R., 195
Kuehn, M. R, 19

L

Labeti, S., 192
Lai, A.-C., 35
Lallier, T., 109
Lance-Jones, C., 213
Landschulz, W. H., 46
Langlet, C., 25
Lassar, A. B., 166
Lauber, B., 37
Leaf, D. S., 147
Leahy, P. S., 8
Lebrach, H., 192
Leder, P., 82
Le Douarin, N. M., 116
Lee, J. J., 8
Lehmann, R., 77
LeMouellic, H., 195
Levine, M., 187, 196
Lewis, J. A, 10
Lieberman, M., 127
Lilien, J., 113
Lim, T. M., 65
Lin, X., 102
Lisowska-Grospierre, B., 48
Lo, D., 117, 167
Loomis, W. F., 5, 87
Lorenz, L., 102
Lowenthal, J. W., 66
Lu, S. D., 122
Luna, E. J., 53
Lunn, E. R., 65
Lye, R. J., 73

M

Mach, B., 48
Maeda, N., 17
Magnuson, T., 16, 19, 20, 22, 23, 45, 50, 69, 83, 122, 170, 171, 175, 177, 199, 219
Mahbulami, H., 203
Malissen, B., 25
Mandel, T. E., 167
Mann, A. M., 129
Manning, S. S., 30
Mardon, G., 170
Marrack, P., 117
Martin, C., 116
Martin, G. R., 194, 198
Mascara, T., 22
Masento, M. S., 202
Matias, P. M., 171
Maxson, R., 35, 38
McClay, D. R., 89, 133
McCune, J. M., 127
McDewitt, H. O., 26
McGee, T., 10
McGhee, J. D., 138, 150
McGillivray, B., 170
McGinnis, W., 197
McIntosh, J. R., 73
McKeown, M., 155, 156
McKinlay, C., 108
McKnight, S. L., 46
McLafferty, S., 10
McLaren, A., 169
McMahon, A. P., 173
McRobbie, S., 203
McVey, J. H., 195
Mege, R.-M., 115
Meier, C., 129
Melton, D. A., 62, 189
Melton, D. W., 17

Menko, A. S., 111
Meper, H. E., 146
Merkl, R., 88
Milburn, M. V., 171
Miller, J. F. A. P., 167
Miller, J. R., 196
Minor, J. E., 8
Mitchell, L., 4
Miura, K., 171
Moerman, D. G., 11, 151
Monk, M., 19
Moon, R. T., 14
Moore, S., 49
Morahan, G., 167
Morris, H. R., 202
Morton, D. G., 71
Mosher, R., 170
Muller, R., 175
Murray, B. A., 115

N

Nagoshi, R. N., 156
Nastiuk, K. L., 162
Neugebauer, K. M., 113
Nguyen-Huu, M. C., 196, 197
Nishimura, S., 171
Nobel-Topham, S. E., 191
Noguchi, S., 171
Noll, M., 76
Norris, M. L., 80
Nottebohm, F., 162
Nusslein-Vollard, C., 75—77

O

O'Farrell, P. H., 42, 181
Ohiki, H., 116
Ohtsucka, E., 171
O'Kane, C. J., 153
Ong, E. S., 220

P

Page, D. C., 170
Palmiter, R. D., 44, 117, 167
Panthier, J. J., 21
Parks, S., 40
Parnes, J. R., 25
Patten, P. A., 119
Perez, C., 122
Peterson, A. C., 81
Petkovich, M., 219
Phillis, R. W., 13
Pines, J., 54
Pintar, J. E., 197
Pollock, J., 94, 170
Por, S. B., 146
Porter, M. E., 73
Pourcel, C., 81
Preston, C. R., 13
Priess, J. R., 71, 136

R

Rabin, B. A., 95
Rabin, L. B., 127
Raff, R. A., 147
Raff, R. R., 55
Raper, J. A., 110
Rathjen, F. G., 110
Regan, C. L., 154
Reichardt, L. F., 113
Reik, W., 45, 80
Reith, W., 48
Renfranz, P. J., 94, 95
Reyes, G. R., 127
Richstein, S., 76
Richter, J. D., 78
Roberts, A. B., 64
Robertson, E. J., 19
Robertson, H. M., 13
Robinson, J. J., 132
Rock, E. P., 119
Rosa, F., 64
Rosbash, M., 101, 102
Rossant, J., 81, 169, 191
Rubin, G. M., 93, 95, 96
Rubin, M. R., 196
Ruddle, F. H., 192, 197
Ruther, U., 175

S

Saari, B., 10
Saez, L., 101
Samraoui, B., 125, 126
Sanchez, C. H., 48
Saper, M. A., 125, 126
Sapienza, C., 81
Sargent, T. D., 15, 44, 61, 62, 64, 79, 80, 104, 189, 190
Satola, S., 48
Sawczuk, I. S., 196
Schaap, P., 52
Schatz, D. G., 23
Schierenberg, E., 92
Schimenti, J., 15
Schindler, J. M., 40, 46, 59, 141, 142, 144, 166, 206, 210, 221

Schmitt-Verholst, A.-M., 25
Schnabel, H., 136
Schnabel, R., 136
Schneuwly, A., 184
Scholey, J. M., 73
Schriefer, L. A., 11, 151
Schroeder, T. E., 206
Scott, M. P., 181
Segui, P., 220
Shackleford, G. M., 173
Shapiro, B. M., 90
Sharpe, C. R., 188
Sharpe, P. T., 196
Sher, E., 42
Showman, R. M., 147
Siekevitz, M., 66
Silver, L. M., 15
Simon, D., 81
Simon, M. A., 96
Simpson, E. M., 170
Sin, C.-H., 86
Sinclair-Thompson, E. Y., 162
Singleton, C. K., 30
Skimming, J., 10
Slawin, K. L., 197
Smith, E., 146
Smith, E. J., 60
Smith, J. C., 60
Smithies, O., 17
Smouse, D., 161
Socolow, D., 15
Soll, D. R., 4
Spangrude, G. J., 142
Spek, W., 52
Sporn, M. B., 64
Spradling, A., 12, 40
Spray, D. C., 101
Springer, T. A., 123
Spudich, J. A., 6
Stacey, A., 22
Steinmetz, J., 120
Stephens, L., 131
Stern, C. D., 65
Sternberg, P. W., 138
Stevenson, M., 129
Steward, R., 208
Stewart, T. A., 82
Stolze, B., 114
Stranten, H. M. W. van, 216
Strome, S., 72
Struminger, J. L., 125, 126
Sucov, H. M., 131, 132
Suda, Y., 68
Superti-Furga, G., 33
Surani, M. A., 45, 80
Suzuke, M., 68
Swain, J. L., 82

T

Taylor, G. W., 202
Technau, G. M., 100
Teitelman, G., 143
Thaller, C., 211
Thoenen, H., 106, 163
Thoma, G., 76
Thomas, J. B., 158, 159
Thomas, K. R., 16
Thompson, S., 17
Thors, F., 216
Timpl, R., 106
Tiollais, P., 81
Todd, J. A., 26
Tomaselli, K. J., 113
Tomlinson, A., 95
Tong, L., 171
Tosney, K. W., 214
Toth, L. E., 196, 197
Tufaro, F., 34

U

Utset, M. F., 197

V

Vardy, P. H., 204
Varmus, H. E., 173
Varnum-Finney, B., 4
Vassin, H., 57
Vigny, M., 104
Vitelli, L., 37
Viviano, C. M., 198
Vold, L., 15
Vos, A. M. de, 171

W

Wagner, E. F., 175
Wallraff, E., 88
Walter, J., 114
Walthall, W. W., 91
Wang, M., 52
Warga, R. M., 140
Wasiak, A., 129
Waterston, R. H., 11, 151
Way, J. C., 180
Weeks, D. L., 62

Weidman, P. J., 90
Weinhart, U., 88
Weintraub, H., 166
Weissman, I. L., 127, 142
Wessel, G. M., 56, 89—91, 133, 134, 147, 207
Westehube, L. J., 53
Wetts, R., 141
White, J. G., 74
White, R. A. H., 181
Widera, G., 167
Wiertz-Hoessels, E. J. L. M., 216
Wildinson, D. G., 173
Wiley, D. C., 125, 126
Williams, J. G., 146, 203
Williams, K. L., 146, 204
Wilt, F., 131, 132
Winter, J., 108
Wolff, J. M., 110
Wolffe, A. P., 43
Wolgemuth, D. J., 198
Wylie, C. C., 79

X

Xu, T., 56

Y

Yaqoob, M., 60
Yoshida, M., 88
Young, M. W., 101, 139
Yu, Q., 102

Z

Zamoyska, R., 25
Zink, B., 192

SUBJECT INDEX

A

Abdominal segmentation, 77
Abdominal signal, 77—78
Abruptex, 58
Accessory proteins, 154
Acetylcholine receptor, 10
N-Acetylglucosaminidase, 29—30
Acidic fibroblast growth factor, 164
Actin, 53—54, 64, 75
Actin-binding proteins, 53
Actin genes, 8—9, 39—40, 149—150
Actin 6 promoter, 5
Actin 15 promoter, 7
Activated T lymphocytes, 48
Adenosine monophosphate, 52—53, 86
Adenoviruses, 129, see also specific types
Aggregation, 6
AIDS, 68, 128—130
Alleles, 80—81, 154—155, see also specific types
Alpha-beta cells, 119
Amanitin, 49
Amino acids, 32, 46, 90, 96, see also specific types
  β chain, 26
  hydrophobic, 99
  of retroviral envelope proteins, 127
  transmembrane, 99
Aminolevamisole, 10
β-Aminopropionitrile (BAPN), 89
Amphibians, 64, see also specific types
Antennae, 184, 186
Antennapedia gene, 181—187
Anterior localization, 75
Anterior patterns, 76—77
Antibodies, 86, 94, 95, 109, 206, see also specific types
  anti-laminin, 108
  to fusion proteins, 147
  to gap junction protein, 205
  generation of, 99
  inhibition of, 105—106
  monoclonal, see Monoclonal antibodies
  polyclonal, 131
Antigen, 25, 66, see also specific types
  cell surface, 81—82, 142, 146
  differentiation, 142
  distribution of, 205
  hepatitis B surface, 81—82
  human leukocyte, see Human leukocyte antigen (HLA)
  identification of, 87
  lymphocyte function-associated, 123—125
  prespore-specific, 146
  recognition of, 126—127
  T cell receptor for, 26
Antigen-presenting cells, 25
Antigen-responsive immunocompetent lymphocytes, 24
Antisense RNA, 5—6, 14—15
Antiserum, 87, see also specific types
Antp gene, 179
Antp induction, 186
Apical-basal polarity, 207
Arrhythmicity, 101—102
Arthritis, 27
Associated processing enzymes, 90, see also specific types
Astrocytes, 113—114
Astroglia, 113
Asymmetric localization, 77
Asymmetry, 72—73
Autoimmune diseases, 27, 167, see also specific types
Autoimmune response, 167
Autonomy, 99—101
Autosomal gene expression, 82
Autosomal transgene, 82—83
Avian, see Birds
AVM, 91, 92
Axis deficiencies, 79—80
Axons, 65, 107—109, 111, 113—115, 162, see also specific types

B

BAPN, see β-Aminopropionitrile
Basal lamina, 104
Basal promoters, 35—37
Basic fibroblast growth factor, 164
B cells, 24, 48, 142, see also specific types
B-chain sequences, 106
BDNF, see Brain-derived neurotrophic factor
BDU cells, 92
Behavioral rhythms, 101
Beta cells, 167—169

β chains, 25, 26
Bicoid activity, 75
Bicoid RNA, 76—77
Biological clock, 102
Birds, 108, 116—117, see also specific types
Blastocytes, 20
Blastomeres, 206—207
Body plan redesign, 184
Body segment development, 179
Bone, 175—177
Bone marrow, 118, 142
Box-binding protein, 48
Brain, 162—163, 175, 191
Brain-derived neurotrophic factor (BDNF), 164—165
B-xyloside, 216

C

Cachectin, 122—123
Caenorhabditis elegans, 9, 10, 71—72
  asymmetry in, 72—73
  ccd-3 mutant and, 137
  cell death in, 137
  cell division axes in, 74—75
  cell interactions in, 92—93
  cellular polarity in, 92—93
  control of gene expression in, 150—151
  cytoplasmic motor in, 73—74
  DNA synthesis in, 150—151
  glp-1 in, 134—137
  meiosis in, 134—135
  mitosis in, 134—135
  neurons in, 91—92
  parental DNA and, 138—139
  touch receptor neurons in, 180—181
  transposable element in, 10—11
  transposon-induced deletions in, 11—12
  tru-1 gene in, 152—153
  unc-22 gene in, 151—152
  vulval cells in, 138
Calcium, 90, 115, 147
Calcium-dependent liver cell adhesion molecules, 115
CAM, see Cell adhesion molecules
Canary, 162—163
Carbohydrates, 87, 89, 91, see also specific types
Carcinoma cell cytoplasmic RNA, 190
CAT, see Chloramphenicol acetyltransferase
Catecholaminergic neurons, 143
N-Catherin, 113—114
CCAAT, 47
CCAAT-binding factor, 33—34
ccd-3 mutants, 137
CD4 cells, 129—130
cDNA, 62, 96, 115—116, 132, 147, 156, 166
CD3 polypeptide complex, 119
Cell adhesion molecules (CAM), 115—116, 120—122, 123—125, see also specific types
Cell-autonomous action, 169—170
Cell-autonomous determination of cell type, 51—52
Cell autonomy, 99—101
Cell-cell adhesion, 85—89, 115—116
Cell-cell binding, 115—116, 121
Cell-cell cohesion, 86
Cell-cell communication, 41, 85, 99, 101
Cell cycle, 51—53, 58—59, 151
Cell death, 137
Cell differentiation, see Differentiation
Cell division, 81
Cell division axes, 74—75
Cell growth, 51
Cell interactions, 85—130, see also specific types
Cell lineage conversion, 133—134
Cell migration, 85
Cell movement, 85
Cell proliferation, 51
Cell sorting, 52—53
Cell specific gene expression, 145—177, see also specific types
Cell surface antigens, 142, 146
Cell surface cytotoxic transmembrane protein, 122—123
Cell surface glycoproteins, 110—111, 146, see also specific types
Cell surface interactions, 147
Cell surface molecules, 201, see also specific types
Cell surface proteins, 147, see also specific types
Cell surface receptors, 112, see also specific types
Cell-type choice, 51—52
Cellular polarity, 92—93
Cellular proteins, 51, see also specific types
Central nervous system, 139, 160, 161, 164, 189
  axons of, 113
  homeobox genes in, 197

neurotrophic factors and, 163—164
Centrosomes, 75
Centrosome separation, 74
C-fos, 175—177
Chemotactic system, 52, 53
Chick
cell-surface glycoprotein in, 110—111
feather patterns in, 215—216
motoneuron projection patterns in, 212—214
motor neurons in, 163—164
neural development in, 65—66
neural tube of, 216—217
neurites in, 214—215
retinoids in, 211—212
Chloramphenicol acetyltransferase, 39, 149
Cholinergic receptor mutants, 10
Chordomesoderm, 189, 217
Chorion genes, 40—41
Chromosomal domains, 45—46
Chromosome 17, 15
Chromosome microdissection methods, 16
C-H-rus p21, 171—173
Ciliogenesis, 148—149, see also specific types
Circadian rhythms, 101—103
Class I histocompatibility molecules, 167—169
Class I major histocompatibility complex, 167—168
Class II major histocompatibility complex, 167
Classical genetics, 3
Cleavage, 71, 72, 80, 127—129
Clock gene, 101
C-myc protein, 146
Coding sequence, 192
Collagen, 22—23, 89—90, 107—110, see also specific types
Collagen genes, 23
Competition, 188
Congenital immunodeficiency, 48—49
Connective tissue, 214
Contact A site protein, 85—86, 88—89
Contact-independent polarization, 206—207
Contraction, 5
Cortex, 206—207
Covaspheres, 86
Cranial neural crest migration, 109—110
C-rel, 208—209
Cross-linking, 89—91
CSAT, 112
Cyclic AMP, 52—53, 86
Cyclin, 54—55
Cycloheximide, 31, 88
Cytochalasin D, 72
Cytodifferentiation, see Differentiation
Cytokinesis, 6, 7
Cytolysis resistance, 129—130
Cytoplasm, 71, 76—78
Cytoplasmic actin genes, 8
Cytoplasmic components, 72
Cytoplasmic determinants, 71—83, see also specific types
Cytoplasmic localization, 71—72, 78
Cytoplasmic motor, 73—74
Cytoplasmic RNA, 190
Cytoplasmic signals, 145
Cytoplasmic surface of ponticulin, 53—54
Cytoplasmic transplantation, 75
Cytoskeleton, 53

D

D19 gene, 146
Dendritic cells, 48
De novo methylation, 82
Depolymeriztion of microtubules, 74
Dermamyotone, 214—215
Dermis, 214
Developmental biology, defined, 1
Developmental gene expression, 29—50
Developmental genetics, 3—28
Developmentally regulated proteins, 32
Developmental programming plasticity, 42
Developmental regulation, 37—38
Developmental timers, 4
Development-controlled enzymes, 29, see also specific types
Diabetes (IDDM), 26—28, 167—169
Dictyostelium discoideum, see also Slime mold
N-acetylglucosaminidase and, 29—30
cell-cell adhesion in, 87—88
cell cycle phase in, 51—53
contact site A protein and, 88—89
DIF and, 203—204
homophilic binding in, 85—86
morphogen differentiation-inducing factor in, 202—203
multicellular slug stage of, 204—205
myosin heavy chain gene and, 5—8
ponticulin and, 53—54
protein synthesis inhibition of gene

expression and, 30—31
structural characterization of, 146
timing mutant of, 4—5
ubiquitin genes of, 32
DIF, 203—204
Differential gene expression, 38—39
Differential 5S RNA gene expression, 43—44
Differential splicing, 120—122
Differentiation, 51, 145—177
cell lineage specific, 147
cell surface, 142
germ cell, 16, 198—199
muscle, 112
myogenic, 111—112
neuronal, 181
posterior neural, 188—189
sexual, see Sexual differentiation
of touch receptor neurons, 180—181
Differentiation-inducing factor, 202—203
Dinucleotides, 83, see also specific types
Diploid cells, 69
Displacement protein, 33—34
DNA
amplification of, 26
elements of in mouse t complex, 15—16
gamete, 139
genetic transmission of, 1
homeobox proteins and, 187—188
methylation of, 81, 82
parental, 138—139
replication of, 81
sequence-specific interaction with, 47
strand segregation in, 139
synthesis of, 26, 150—151
transfer of, 37
DNA binding proteins, 33, 46—48, 179, 187, see also specific types
DNA binding specificity, 47
DNA-protein complex, 48
Dorsal ectoderm, 189
Dorsal gene, 208—209
Dorsal root ganglia, 65
Dorsoventral (DV) axis, 207—208
Dorsoventral (DV) polarity, 210
Drosophila melanogaster, 13—14
neurogenic gene delta of, 57—58
neutogenic genes of, 99—101
notch locus of, 56—57
nucleotides of, 96—99
period gene in, 102—103
sexual differentiation in, 155—156
Drosophila sp., 12—13
abdominal signal in, 77—78
antennapedia gene in, 185—187
anterior pattern of, 76—77
bicoid activity in, 75
chorion genes and, 40—41
clock gene of, 101
DNA in, 187—188
EGF-like coding sequences in, 139—140
engrailed gene in, 42—43, 194—195
eye of, 94—96
genomic regulatory elements in, 153—154
homeotic genes in, 181—184
inducible promoters and, 101—102
nervous system development in, 158—159
neurogenesis in, 160—161
neuronal fate in, 161—162
neuronal pattern formation in eye of, 94
polarity gene in, 208—209
redesigning of body plan of, 184
segmentation genes in, 181—184
sevenless gene of, 93—94
toll gene of, 210
transformer gene in, 156—158
β-tubulin gene of, 154—155
DV, see Dorsoventral
Dynein, 74
Dynetin, 73

E

Early histone genes, 37—39
E cells, 52
ECM, see Extracellular matrix
Ecotropic murine leukemia virus, 21—22
Ectoderm, 139, 189
Ectoderm cells, 104, 150
Ectodermis, 148—149
EDTA-resistant binding, 86, 87
EGF, see Epidermal growth factor
EHS, see Engelbreth-Holm-Swarm
Embryology, 1
Embryonic lens induction, 103—104
Embryonic stem cells, 17—19
En-1 gene, 194
En-2 gene, 191, 194
Endocrine cells, 143—144
Endoderm, 62

Endoproteolytic cleavage, 127
Endothelial cells, 123—125
Engelbreth-Holm-Swarn (EHS) tumor, 106
En gene, 179
Engrailed gene, 42—43, 170—184, 194—195
Enhancers, 29, 46, 81, see also specific types
Entactin/nidogen, 105
Enveloped viruses, 128, see also specific types
Envelope glycoproteins, 129—130, see also specific types
Enzymes, 19, 29, 30, 90, see also specific types
Epidermal cells, 58, 100
Epidermal growth factor (EGF), 55—56, 58, 139
Epidermal growth factor (EGF)-like coding sequences, 139—140
Epidermal growth factor (EGF)-like proteins, 55—57
Epidermal growth factor (EGF)-like repeats, 57—58
Epidermal pathways, 101
Epidermis, 104
Epithelial grafts, 116—117
Escherichia coli, 153
N-Ethyl maleimide, 74
Ethylmethane sulfonate, 11
Eukaryotes, 1, 29
Even-skipped gene, 161—162
Evolution, 1
Exons, 44, see also specific types
Extracellular EGF-like domain, 56—57
Extracellular matrix, 22, 89—91, 111—112
Eye, 94, 95, 99, see also specific parts

F

F-actin, 53—54
Fate-determining mechanism, 138
Feather pattern formation, 215—216
Fertilization envelope, 90—91
FGF, see Fibroblast growth factor
Fibril formation, 23
Fibroblast growth factor (FGF), 61—62, 164
Fibroblasts, 23—25, 166
Fibronectin, 107—109, 112
Fibronectin/laminin receptor, 112
Finger protein, 170—171
Floor plate area, 216—217
FM-1, 4
Forebrain, 162, 163
Foreign antigen binding sites, 126—127
Frog, 1, 141—142
Fruit fly, 1
fstA1, 5
Full length proteins, 187
Fushi tarazu, 160—161, 181—184
Fusion, 39—40, 153
Fusion genes, 149—150
Fusion proteins, 147, see also specific types

G

Gamma-delta cells, 119—120
β-Galactosidase, 103, 153
Gamete DNA, 139
Gametogenesis, 80
Ganglia, 65, see also specific types
Ganglion cells, 108
Gap genes, 182
Gap junctional communication, 205—206
Gap junction proteins, 205, 206
Gastrulation, 89—90, 159, 193, 217
GDP, see Guanosine diphosphate
Gene expression, 29—50, see also specific types
  autosomal, 82
  basal promoters and, 35—37
  cell specific, 145—177, see also specific types
  deactivation of, 31
  differential, 38—39
  differential 5S RNA, 43—44
  DNA synthesis and control of, 150—151
  hepatitis B surface antigen, 81—82
  human, 49—50
  myosin heavy chain, 5—6
  ontogenic promoters and, 35—37
  post-transcriptional restriction of, 34—35
  protein synthesis inhibition of, 30—31
  region-specific, 197—198
  regulation of, 179
  spatial, 29, 40, 45, 102—103
  spatially deranged, 39—40
  spatially patterned, 42—43
  spatially regulated, 40—41
  temporal, 29, 45, 102—103, 174
  temporally correct, 39—40
  timing of, 35—37

tissue-specific, 145—177, see also specific types of reporter gene, 154
Gene products, 77, 201, see also specific types
Genes, see also specific types
actin, see Actin genes
chorion, 40—41
clock, 101
collagen, 23
cytoplasmic localization and, 71—72
disruption of, 6
fusion, 149—150
gap, 182
heat shock, 184
histone, see Histone genes
homeobox, see Homeobox genes
homeotic, 161, 181—184
hybrid insulin, 143—144
lineage-specific, 131—133
maternal, 75
neomycin resistance, 17
nervous system structure and, 9
neurogenic, 99—101
pair-rule, 182
polarity, 42, 43, 208—209
realization, 181
regulation of, 1, 38
regulatory, 29
reporter, 154
segmentation, see Segmentation genes
segment polarity, 182
sex determining, 152—153, 158
structural, 29, 40
T-cell receptor, 23
testis-determining, 169—170
transfer of, 23—25, 24
transformation of, 6
transformer, 155—158
ubiquitin, 32
zygotic, 182
Gene targeting, 6, 16—18
Genetics, 1
classical, 3
developmental, 3—28
Genetic segregation, 8—9
Genetic transmission of DNA, 1
Genomes, 1, 29, 35, 43, 148, see also specific types
Genomic regulatory elements, 153—154, see also specific types
Germaline, 19
Germ cells, 16, 173—175, 198—199
Germ line-specific P granules, 72
Glia, 106
Glial Muller cells, 141
Glp-1, 134—137
Glucocorticoids, 177, see also specific types
Glycine, 22
Glycocalyx, 90
Glycoproteins, 48, 107, see also specific types
cell-surface, 110—111, 146
contact A site, 85—86
envelope, 129—130
17-kD integral, 53—54
membrane, 86, 120
surface, 87—88
Glycosylation, 87—89, 216
Gonad, 169
Gonadal tissue, 145
Gonadogenesis, 136
Gp24, 87—88
G proteins, 171
Growth factors, 51, 56, 61—64, 145, 164—165, see also specific types
GTP, see Guanosine triphosphate
Guanosine diphosphate, 172
Guanosine triphosphatase, 172, 173
Guanosine triphosphate (GTP), 171, 172

H

Haploid genome, 148
Haplotypes, 15, 16
Haywire, 154—155
Head, 185—187
Heart, 175
Heat-inducible promoter, 184
Heat shock, 186
Heat shock gene, 184
Heat shock promoter, 185
Heavy-chain enhancer, 81
Heavy meromysin (HMM), 6, 7
Heavy metals, 177, see also specific types
Helix-turn-helix, 47
Helper T cells, 129, 130
Hematolymphoid system, 142
Hematopoietic stem cells, 142—143
Heparan sulfate proteoglycans (HSPG), 104—105, 109—110
Hepatitis B, 81—82
Hepatitis B surface antigen, 82

Hermaphrodites, 152, 169
Herpes simplex virus, 46
Heteromeric protein complexes, 58
H genes, 31, 35—37
High frequency switching, 4—5
High-molecular-weight proteins, 73
Hindlimb, 212—214
Histocompatibility molecules, 167—169
Histogenesis, 115
Histone gene promoter, 33—34
Histone genes, 35—39
HIV, see Human immunodeficiency virus
HLA, see Human leukocyte antigen
HMM, see Heavy meromysin
HNK-1, 65
Holzer's quantal cell cycle model, 151
Homeobox genes, 18, 162, 179—199, see also specific types
  conserved nature of, 198
  developmental role of, 191
  differential expression of, 198—199
  en-2, 191
  engrailed class, 179—180
  expression of, 196—197
  hox 1.3, 191—193
  hox 1.4, 198—199
  hox 1.5, 192, 193
  hox 1.6, 190—191
  hox 1.7, 196
  hox 2.1, 194, 195
  hox 3.1, 195—196
  isolation of, 196—197
  mammalian, 191
  mec-3, 180—181
  region-specific expression of, 197—198
Homeodomain-containing proteins, 187
Homeotic genes, 93—94, 161, 181—184, see also specific types
Homologous recombination, 6—8, 18, 19
Homophilic binding, 85—86
Homophilic cell-cell bonds, 121
Hox 1.3 gene, 191—193
Hox 1.4 gene, 198—199
Hox 1.5 gene, 192, 193
Hox 1.6 gene, 190—191
Hox 1.7 gene, 196
Hox 2.1 gene, 194, 195
Hox 3.1 gene, 195—196
HPRT, see Hypoxanthine-guanine phosphoribosyl transferase
HSPG, see Heparan sulfate proteoglycans
Human C-H-rus, 171—173
Human embryo, 49
Human gene expression, 49—50
Human immunodeficiency virus (HIV), 66, 68, 127—130
Human leukocyte antigen (HLA)-A2, 125—127
Human leukocyte antigen (HLA)-D, 27
Human leukocyte antigen (HLA)-DQ gene, 26—28
Human leukocyte antigen (HLA)-DR promoter binding protein, 48—49
Human Y chromosome, 170—171
Hybrid insulin genes, 143—144
Hydra, 205—206
Hydrophobic amino acids, 99
Hyperglycemia, 167
Hypoxanthine-guanine phosphoribosyl transferase (HPRT), 17—20
Hypoxanthine phosphoribosyl transferase, 17

I

ICAM, see Intercellular CAM
IDDM, see Insulin-dependent diabetes mellitus
Immune responses, 68
Immune system, 85, 145
Immunocompetent lymphocytes, 24
Immunocompetent T cells, 66, 118
Immunodeficiency, 48—49, see also specific types
Immunoglobulin, 23—25
Immunologic recognition of self, 116—117
Imprinting, 71—83
  in mammals, 83
  methylation and, 80—81
  molecular mechanism for, 82—83
Inducible promoters, 101—102
Inducible proteins, 66, see also specific types
Inflammatory response, 167
Influenza, 128
Inheritance evolution, 1
Insertional mutagenesis, 12—13
Insulin, 143—144
Insulin-dependent diabetes mellitus (IDDM), see Diabetes
Int-1, 173—175
Integrins, 111—114
Intercellular CAM, 123—125
Interference, 129

Interleukin-1, 123
Interleukin-2, 66—68
Intracellular junctional communication, 101
Intracellular localization, 72
Intrinsic circadian oscillator, 102
Introns, 44—45, see also specific types
Islet cells, 143

J

Junctional communication, 101
Jurkat T cells, 66, 68

K

Kidney, 175
Kinesin, 74

L

Laminin, 104—109, 113
Laminin/fibronectin receptor, 108—109
Laminin-heparan sulfate proteoglycan complex, 109—110
Late-developing neurons, 91—92
Late histone genes, 37—39
L-CAM, see Liver CAM
L cells, 52
Legs, 184, 185
Lens induction, 103—104
Lesch-Nylan syndrome, 19—20
Leucine zipper, 46—47
Leukemia virus, 21—22
Levamisole, 10
LFA, see Lymphocyte function-associated antigen
Limb bud, 211—212
Lineage-specific genes, 131—133
Line deficiencies, 79—80
Linkage analysis, 3, 8
Linkage patterns, 8
Lipopolysaccharides, 122
Liver CAM (L-CAM), 115, 116
Liver cell adhesion, 115—116
Localization, 72, 75—78, 95, 99, 151—152, see also specific types
Long terminal repeat (LTR), 66
Lumbosacral level, 214—215
Lung, 175
Lymphocyte function-associated antigen-1 (LFA-1), 123—125
Lymphocytes, 24, 123—125, see also specific types
Lymphoid system, 124, 142
Lysosomal enzymes, 29, see also specific types
Lyt-2 gene, 25—26

M

M46 protein, 59
M116 protein, 59
Macrophages, 48, 123
M actin gene, 8
Magnesium ATPase, 73
Major histocompatibility complex (MHC), 25—27, 48—49, 167—168
Mammalian growth factors, 51, 61, 62, see also specific types
Mammalian homeobox genes, 191
Mammals, 83, see also specific types
Maternal controls, 71—83, see also specific types
Maternal genes, 75
Maternal genome, 35
Maternal mRNA, 62—63
Matrix proteins, 131—133, see also specific types
Maturation promoting factor (MPF), 54, 59
Mecamylamine, 10
Mec-3 gene, 180—181
Meiosis, 134—135
Membrane protein receptor, 112
Membrane proteins, 14, see also specific types
Membranes, 86, 95—96, 114—115, 120, see also specific types
Membrane skeleton protein 4.1, 14—15
Mesenchyme cells, 133—134, see also specific types
Mesenchyme-specific cell surface proteins, 147
Mesoderm, 191—192, 197, 217
  inducers of, 62
  induction of, 64
  morphogenesis of, 133
  synergistic induction of, 61—62
Mesodermal pattern, 60—61
Mesoderm-inducing factor (MIF), 60—61
Meta-aminodevamisole, 10
Metabolism, 20, see also specific types
Metallothionein, 45

Metallothionein promoter, 177
S-Methionine, 49
Methylation, 80—83
MHC, see Major histocompatibility complex
Microfilament inhibitors, 72, see also specific types
Microfilaments, 72—73
Microinjection, 22, 23, 37—38, 54, 189—190
Microtubule-associated proteins, 155, see also specific types
Microtubule binding proteins, 73, see also specific types
Microtubules, 73—74, 154—155
MIF, see Mesoderm-inducing factor
Migration, 109—110
Mitogen, 66—68
Mitosis, 54, 134—135
Molecular biology, 3
Molecular characterization, 1
Molecular cloning, 54—55
Molecular events, 29, see also specific types
Molecular genetics, 1
Monoclonal antibodies, 30, 65, 86—88, 95, 96, 108, 112, 142, 201, 202, 205, see also specific types
  laminin-heparan sulfate proteoglycan and, 109—110
  to LFA, 123—125
  MUD 1, 149
  RNA-binding protein and, 79
Monocyte-macrophages, 48, 123
Monocytes, 122
Monokines, 123
Monosaccharides, 91, see also specific types
Morphogen, 203—204, 220—221, see also specific types
Morphogen differentiation-inducing factor, 202—203
Morphogenesis, 6, 201—221
  CAMs in, 115
  ECM and, 90
  mesoderm, 133
  sea urchin, 89
Morphogenetic molecules, 201
Mosaic analysis, 3, 9
Motoneuron projection patterns, 212—214
Motor axons, 65
Motor neurons, 163—164
Mouse, 1
  bone development in, 175—177
  brain of, 191
  chromosomal domains in, 45—46
  diabetes in, 167—169
  differential splicing in, 120—122
  embryo-derived stem cells in, 16—17
  en gene in, 194
  gastrulation and, 193
  genomic imprinting and, 80—81
  hematopoietic stem cells in, 142—143
  hox 1.6 in, 190—191
  hox 2.1 in, 194
  Lesch-Nyhan, 19
  N-CAM and, 120—122
  polyadenylation in, 120—122
  proto-oncogene in, 173
  stem cells in, 68—69
  SWR/J, 21—22
  t complex of, 15—16
  transcriptional efficiency in, 44—45
  XY chimaeric, 169—170
MPF, see Maturation promoting factor
mRNA, 32, 35, 44, 79, 189—190
  accumulation of, 146
  alpha-actin, 64
  chorion protein, 40
  coding of for FGF, 61—66
  cyclin and, 54—55
  maternal, 62—63
  reversal of, 162—163
  synthesis of, 80
  synthetic, 54
  tra, 156
  unfractionated, 61
MUD 1, 146
Muller cells, 141
Multicellular slug stage of *Dictyostelium discoideum*, 204—205
Multipotent precursors, 141—142
Murine leukemia virus, 21—22
Muscle, 5, 112, 163—164, 175, 214, see also specific types
Muscle heavy meromysin (HMM), 6, 7
Mutagenesis, 12, 16—17, 76
Mutant phenotypes, 3, see also specific types
Mutants, 4—5, 10, 17—19, see also specific types
Myelomonocytic cells, 142
Myoblasts, 112, 166
Myogenic differentiation, 111—112
Myosin, 5, 6, 151

Myosin heavy chain gene, 5—8

N

Nasal axons, 114
Nc2 allele, 154—155
N-CAM, see Neural CAM
Nematodes, 10, see also specific types
Neomycin resistance, 17
Nerve growth factor (NGF), 164—165
Nerves, 92, 158, 214, see also specific types
Nervous system, 85, 145, 154, 181, see also specific types
cell types in, 163
central, see Central nervous system
development of, 107, 137, 158—159
peripheral, 65, 165
regeneration of, 107
structure of, 9
Neural CAM (N-CAM), 115, 116, 120—122
Neural cells, 115—116, 173
Neural crest cells, 65, 109—110
Neural development, 65—66
Neural differentiation, 188—189
Neural induction, 188—189
Neural retinal, 141
Neural tube, 173—175, 216—217
Neurites, 105, 106, see also specific types
elongation of, 107
interactions among, 110—111
outgrowth of, 104—106, 109—110, 113, 214—215
patterned, 214—215
sensory, 107—108
Neurofuscin, 110—111
Neurogenesis, 160—161
Neurogenic element, 161
Neurogenic genes, 57—58, 100
Neurogenic loci, 57
Neurogenic pathways, 101
Neurogenic protein, 139
Neuronal cells, 100
Neuronal connections, 113
Neuronal fate, 161—162
Neuronal guidance, 113
Neuronal pattern formation, 94
Neuronal receptors, 113
Neurons, 106, 162, see also specific types
astroglia and, 113
catecholaminergic, 143
in ccd-3 mutants, 137
death of, 164—165
differentiation of, 181
hybrid insulin genes and, 143—144
late-developing, 91—92
motor, 163—164
outgrowth of, 113—114
pharyngeal, 137
spinal cord, 104
spinal motor, 163—164
survival-promoting of, 164
touch receptor, 180—181
Neurotoxic drugs, 10, see also specific types
Neurotrophic proteins, 164, see also specific types
Neutogenic genes, 99—101
Neutroophic factors, 163—165
NGF, see Nerve growth factor
N-linked carbohydrates, 89
Nonantigenic B-chain sequences, 106
Nonmuscle cells, 5
Notch locus, 56—57, 139—140
Notochord, 216—217
Nuclear material, 58—59, see also specific types
Nuclear proteins, 66—68, 159—160, see also specific types
Nuclear receptors, 219—220, see also specific types
Nucleic acids, 218, see also specific types
Nucleotides, 96—99, see also specific types

O

Occipital somites, 65—66
Octamer, 33
Oligosaccharides, 89, see also specific types
Oncogene protein, 171—173
Oncogenes, 18, 145, 171—175, 201, 208—209
Ontogenic promoters, 35—37
Oocytes, 78—80
Oogenesis, 40—41
Operon fusion, 153
Optic pathway, 108
Optic tract, 109
Optic vesicle, 103—104
Organelles, 154, see also specific types
Organ function, 124
Osteogenesis imperfecta, 22—23
Ovoperoxidase, 90, 91

P

Pacemakers, 103
Pair-rule genes, 182
Pancreas, 175
Pancreatic beta cells, 167—169
Pancreatic endocrine cells, 143—144
Pancreatic islet cells, 143
Parental alleles, 80—81
Parental DNA, 138—139
Parental imprinting, see Imprinting
Partial fertilization envelope, 90—91
Patterned substrata, 107—108
Patterns, see also specific types
  feather, 215—216
  formation of, 201—221
  gap junctional communication disruption and, 205—206
  in hydra, 205—206
  motoneuron projection, 212—214
  spatial, 201—202
Pax 1, 218—219
P element-mutagenesis method, 12
P elements, 12—14
Peptides, 51, see also specific types
Per action, 103
Per gene, 101, 159—160
Perinatal lethal osteogenesis imperfecta, 22—23
Period, 101—103
Peripheral nervous system, 65, 165
Per-mutant arrhythmicity, 101—102
P granules, 72, 73, 75
Pharyngeal cells, 136, 137
Phenotypes, 3, see also specific types
Phenotypic switching, 5
Phosphatase, 54
Phosphodiesterase, 52
Phosphorylation, 59, see also specific types
Photoreceptor cells, 96
Photoreceptor R7, 95—96
Plasma membranes, 14, 53—54
Plasticity in developmental programming, 42
PLML, 9
PLMR, 9
PMA, 122
PMC, see Primary mesenchyme cells
Point mutations, 56—57
Polarity, 92—93, 207, 210, 212
Polarity genes, 42, 43, 208—209
Polarization, 206—207, see also specific types
Polyadenylation, 120—122, 162
Polyclonal antibodies, 131, see also specific types
Polymerization inhibitors, 74
Polymorphisms, 8, see also specific types
Polypeptides, 25, 49, 55—56, 62, 87, see also specific types
Polyphondylium pallidum, 201—202
Ponticulin, 53—54
Positional effects, 46
Positional information, 95, 206
Position-specific properties, 114—115
Posterior neural differentiation, 188—189
Post-transcriptional restriction of gene expression, 34—35
Post-translational glycosylation, 87—89
Precursors, 141—142, see also specific types
Predetermination, 188—189
Preimplantation development, 49—50
Prespore-specific surface antigen, 146
Primary mesenchyme cells (PMC), 133—134
Primary transcription factor, 44
Pro-α1, 22—23
Programmed cell death, 137
Prokaryotes, 153, see also specific types
Proliferation, 51, 68—69, 171
Proline, 90
Promoters, 29, 46, see also specific types
  actin 6, 5
  actin 15, 7
  heat-inducible, 184
  heat shock, 185
  histone gene, 33—34
  inducible, 101—102
  metallothionein, 177
  neurogenic element from, 161
  thymidine-kinase, 46
  tumor, 122
Protein phosphatase, 54
Proteins, see also specific types
  accessory, 154
  actin-binding, 53
  box-bining, 48
  cell-cell adhesion and, 88
  cell surface cytotoxic transmembrane, 122—123
  cellular, 51
  chemistry of, 1
  contact site A, 88—89
  developmentally regulated, 32
  displacement, 33—34
  DNA binding, 46—48, 179, 187

epidermal growth factor-like, 55
extracellular EGF-like domain of, 56—57
finger, 170—171
full length, 187
fusion, 147
G, 171
gap junction, 205, 206
heteromeric, 58
high-molecular-weight, 73
homeodomain-containing, 187
inducible, 66
localization of, 99
matrix, 131—133
membrane, 14
membrane skeleton, 14—15
mesenchyme-specific cell surface, 147
microtubule-associated, 155
microtubule binding, 73
neurogenic, 139
neurotrophic, 164
nuclear, 66—68, 159—160
oncogene, 171—173
proteoglycan core, 216
putative transmembrane, 57—58
recombinant sim, 160
regulatory, 47
retroviral envelope, 127
RNA-binding, 78—79
species-specific, 34, 35
spicule matrix, 131, 132
structural, 47
synthesis inhibition of, 30—31, 88
transmembrane, 122—123, 210
viral envelope, 129
Proteoglycan, 104—105, 215—216
Proteoglycan core proteins, 216
Proteoliasin, 90, 91
Proteolysis, 88
Proto-oncogenes, 18, 145, 173—175, 201, 208—209
Proviruses, 21, see also specific types
Proximal tissues, 214—215
PsA glycoprotein, 146
Psammechinus miliaris, 37
Pseudo cells, 137
Pumillio gene, 77—78
Purine, 20
Purine salvage enzyme, 19
Putative antigen recognition groove, 126
Putative cytoplasmic determinants, 75
Putative receptors, 95
Putative transmembrane protein, 57—58
Putative transmembrane receptor, 93—94

Q

Quantal cell cycle, 151

R

RA, see Retinoic acid
Rapid phenotypic switching, 5
RAR, see Retinoic acid receptor
Rate-limiting components, 4
R7 cells, 95—96
Realization genes, 181
Receptors, see also specific types
acetylcholine, 10
cell surface, 112
cholinergic, 10
extracellular matrix, 111—112
fibronectin/laminin, 112
laminin/fibronectin, 108—109
membrane protein, 112
for morphogen retinoic acid, 220—221
neuronal, 113
neuron-astroglia interactions and, 113
nuclear, 219—220
putative, 93—94, 95
retinoic acid (RAR), 219—220
T cell, see T cell receptors (TCR)
touch, 180—181
transmembrane, 93—94
tyrosine kinase, 96
Recombinant sim protein, 160
Recombination, 6—8, 18, 19, 23, 24
Redundancy of function, 11
Regeneration, 115
Region-specific gene expression, 197—198
Regulation, see also specific types
developmental, 37—38
in earliest phase of development, 38
of ECM, 90—91
of gene expression, 179
spatial, 41, 173
temporal, 173
of transcription, 47
two-tiered, 42—43
upstream, 43
Regulatory cascade, 42
Regulatory defects, 48—49
Regulatory genes, 29
Regulatory mechanisms, 38

Regulatory proteins, 47
Reporter genes, 154, see also specific types
Resistance, 10, 17, 26—28, 129—130, see also specific types
Restriction fragment length polymorphisms (RFLP), 8
Retina, 95—96, 114, 141—142
Retinal axons, 109, 111, 114—115
Retinal ganglion cell (RGC) axons, 108—109
Retinoic acid, 211, 212, 220—221
Retinoic acid receptor (RAR), 219—220
Retinoids, 211—212
Retinol, 219
Retino-tectal map, 114
Retroviral envelope proteins, 127, see also specific types
Retroviral recombination, 24
Retroviral vectors, 129
RFLP, see Restriction fragment length polymorphisms
RF-X, 48—49
RGC, see Retinal ganglion cell
Rheumatoid arthritis, 27
Rhythmic behavior, 102
Rhythmicity, 102
RN22 Schwannoma cells, 106
RNA, 76—77, 155—156, 162, 190
RNA-binding proteins, 78—79
RNAse, 162
Rous sarcoma virus (RSV), 82

S

Salivary glands, 103, 175
Schwannoma cells, 106
SCID, see Severe combined immunodeficiency disease
SCW, see Surface contraction waves
Sea urchin, 1, 34
  actin gene in, 8—9, 39—40
  blastomeres of, 206—207
  cell lineage conversion in, 133—134
  dorsoventral axis in, 207—208
  EGF in, 55—56
  fusion genes in, 149—150
  gastrulation in, 89—90
  growth factors in, 56
  H1 gene expression in, 35—37
  histone genes in, 37—39
  homeobox genes in, 179—180
  interspecies hybrid embryos of, 34—35
  matrix protein of, 131—133
  mesenchyme-specific cell surface protein of, 147
  morphogenesis in, 89
  mRNA for cyclin from, 54—55
  post-transcriptional restriction of gene expression in, 34—35
  spec3 in, 148—149
Secondary mesenchyme cells (SMC), 133—134
Segmental identity, 186
Segmentation genes, 77, 160—162, 181—184, see also specific types
Segmented structures, 218—219
Segment polarity genes, 182
Self-assembly, 91
Self-MHC restriction, 118, 119
Self recognition, 116—117
Sendai virus, 128
Sensory axons, 107
Sensory ganglia, 65
Sensory neurites, 107—108
Sequence-specific interaction with DNA, 47
Sertoli cells, 169—170
Sevenless, 93—94
  in cell membranes, 95—96
  localization of, 95
  molecular characterization of, 94
  nucleotides of, 96—99
Severe combined immunodeficiency disease (SCID), 48
Sex combs reduced gene, 181—184
Sex determination, 145, 156
Sex determining genes, 152—153, 158, see also specific types
Sex-determining region of chromosomes, 170—171
Sexual differentiation, 155—157
Sexual transformations, 152
Shock, 184—186
Side branch formation, 202
Single-minded locus, 158—159
Single P elements, 12—13
Skeletal muscle, 163—164, 214
Skeleton protein 4.1, 14—15
Slime mold, 29—30, 51, 201—202, see also specific types
SMC, see Secondary mesenchyme cells
SM50 protein, 132
Somatic cells, 68
Somatic recombination, 23
Somatic sexual differentiation, 157

Somites, 65—66, 189—190, 212—214, see also specific types
Spatial arrangement of molecules, 79
Spatial gene expression, 29, 40, 45, 102—103
Spatially deranged gene expression, 39—40
Spatially patterned gene expression, 42—43
Spatial pattern, 201—202
Spatial regulation, 41, 173
Spec3, 148—149
Species-specific proteins, 34, 35, see also specific types
Specificity, 47, see also specific types
Spermatogenesis, 174
Spicule matrix proteins, 131, 132
Spicules, 131—133
Spiculogenesis, 90, 147
Spinal cells, 104, 163—164
Spinal nerves, 214
Stem cells, 17—19, 68—69, 142—143
Strongylocentrotus purpuratus, see Sea urchin
Structural enzymes, 30
Structural features, 29, see also specific types
Structural genes, 29, 40
Structural proteins, 47, see also specific types
Submembranous network, 53
Substrata, 107—108, see also specific types
Substrate-absorbed laminin, 107
Surface contraction waves (SCW), 59
Surface glycoprotein, 87—88
Switching, 4—5
SWR/J mouse, 21—22
Synthetic mRNA, 54
Synthetic Xhox-1A, 189—190

T

3T3 fibroblasts, 23—25
Targetted correction of mutants, 17—19
TATA box, 33, 48
Tc1, 11
T cell receptors (TCR), 23, 25—26, 117, 119—120
T cells, 142, see also specific types
  activation of, 48, 68
  antigen-responsive immunocompetent, 24
  development of, 117—119
  diabetes and, 168
  growth of, 66—67
  helper, 129, 130
  immunocompetent, 66, 118
  Jurkat, 66, 68
  maturation of, 117
  recognition of, 120, 126—127
  selection of, 118
  thymus environment and, 117—119
TCR, see T cell receptor
TDF, see Testis-determining factor
Tectal cell membranes, 114—115
Tectum, 114, 115
Temporal axons, 114, 115
Temporal gene expression, 29, 45, 102—103, 174
Temporally correct gene expression, 39—40
Temporal regulation, 173
Teratocarcinoma, 19
Testis-determining factor, 170, 171
Testis-determining gene, 169—170
Testis-specific β-tubulin gene, 154—155
TF, see Transcription factor
TGF, see Transforming growth factor
TGFβ, see Transforming growth factor
6-Thioguanine, 17
Thoracic segments, 184
Thoracic transformations, 185—187
Thorax, 187
Thymic epithelial grafts, 116—117
Thymidine-kinase promoter, 46
Thymocytes, 118—120
Thymus, 116—119
Timers, 4
Timing mutants, 4—5
Tissue specific gene expression, 145—177, see also specific types
TNF, see Tumor necrosis factor
Tolerance, 116—119, 167, see also specific types
Toll gene, 210
Touch receptor neurons, 180—181
Tra-1 gene, 152—153
Tra mRNA, 156
Trans-acting factors, 43, see also specific types
Transcription, 181, 188, 189
  accumulation of, 36
  efficiency of, 44—45
  inhibitors of, 49
  patterns of, 156
  rate of, 31
  regulation of, 47

Transcription factor, 44, 48
Transformer gene, 155—158
Transforming growth factor beta (TGFβ), 62
Transforming growth factor (TGF), 62, 64
Transgenes, 45—46, 82—83
Transgenic embryos, 39—40
Translational control mechanisms, 79
Transmembrane amino acids, 99
Transmembrane anchor, 96
Transmembrane proteins, 57—58, 122—123, 210, see also specific types
Transplantation, 75, 77, 78
Transposable elements, 10—11
Transposase, 12, 13—14
Transposition, 10
Transposons, 11—12, 151, see also specific types
Troponin I, 81
Trunk somites, 65—66
Trypsin, 86
Tubulin, 155
β-Tubulin gene, 154—155
Tunicamycin, 88
Tumor cell metastasis, 124
Tumor necrosis factor (TNF), 122—123
Tumor promoters, 122, see also specific types
Tumors, 106, 171, 173, see also specific types
Two-tiered regulation, 42—43
Type I collagen, 22
Type IV collagen, 104, 110
Tyrosine kinase, 93—94
Tyrosine kinase receptors, 96

U

Ubiquitin genes, 32
Ultrabithorax proteins, 181—184
Ultraviolet irradiation, 79—80
Unc-22 gene, 11—12, 151—152
Unfractionated mRNA, 61
Upstream regulatory region, 43
Upstream sequence elements, 36
USE IV, 36

V

Vegetal hemisphere, 62—63
Ventral ectoderm, 189
Viral envelope proteins, 129
Viruses, 21—22, 46, 81—82, 128, see also specific types
Vitamin A, 219
Vitelline layer, 90, 91
Vulval cells, 138

X

Xenopus laevis, 43, 60—61
Xenopus sp., 14—15
  cell cycle of, 58—59
  FGF in, 61—62
  growth factors and, 62—63
  RNA-binding protein and, 78—79
  somite formation in, 189—190
  UV irradiation and, 79—80
Xenopus Tissue Culture (XTC) cells, 60, 61, 64
Xhox-1A, 189—190
XY chimaeric mouse, 169—170
Xyloside, 216

Y

Y chromosome, 170—171
Yeast, 6, see also specific types

Z

Zebrafish, 140—141
Zinc-coordinating repeating units, 47
Zone of polarizing activity (ZPA), 212
Zygotic genes, 182